APPLIED RESEARCH FOR CORN PRODUCTION
IN INDIANA, 2024

APPLIED RESEARCH FOR CORN PRODUCTION IN INDIANA, 2024

DANIEL QUINN

PURDUE UNIVERSITY PRESS
WEST LAFAYETTE, INDIANA

Cataloging-in-Publication Data on file at the Library of Congress.

978-1-62671-238-6 (paperback)
978-1-62671-239-3 (epdf)

Cover image: Purdue University Department of Agriculture

CONTENTS

ACKNOWLEDGMENTS

This report entails a detailed summary of applied field research trials for corn production systems in Indiana, conducted under the direction of Dr. Daniel Quinn and the Purdue Corn Agronomy team in the Department of Agronomy at Purdue University. The authors extend many thanks to the Purdue Agronomy Center for Research and Education, the Purdue Agricultural Centers, farmer cooperators, and the many industry collaborators and funding agencies who help provide the necessary resources needed to support this research. Special recognition is extended to Betsy Bower, Erick Oliva, Emely Gramajo, Narciso Zapata, and Jose Vaca who assisted with trial organization, data collection and processing, and the preparation of this report. In addition, the authors also extend thanks to Crystal Paris for report booklet design and to visiting scholars and undergraduate students Victor Hernandez, Galo Vera, Johnny Garrido, Josh Terrell, and Quinn Gerber who assisted with trial organization, data collection, and scouting. Overall, the combined efforts of various colleagues, professionals, students, and farmers are responsible for the success of this research.

The authors would also like to thank the following for their support in 2024:

Indiana Corn Marketing Council
Corteva Agriscience
Arclin
Bayer Crop Science
FMC
Valent
UKT Chicago Inc.
BASF
Brandt
National Science Foundation
The Popcorn Board
K + S North America

USDA-NIFA
John Deere
Netafim
Cargill
Purdue University
Mosaic
Becks Hybrids
Koch Agronomic Services
Pivot Bio
Nutrien
Teva Corp.
Azotic Technologies

SUMMARY OF THE 2024 CORN GROWING SEASON IN INDIANA

In 2024, Indiana's statewide average corn yield was 198 bushels per acre (bu/ac; Figure 1), falling short of the 2023 record of 203 bu/ac. However, the state average yield was still above trend by ~6%. The season began with favorable planting conditions and near-normal planting progress, with 54% of the crop planted by May 19—just three percentage points below the five-year average. These conditions supported timely crop emergence, strong early plant establishment, and favorable crop condition ratings heading into late May and June. However, dry conditions emerged during this period, mirroring those of 2023. For instance, Purdue University's research farms in West Lafayette, IN, and Butlerville, IN, received only 2.8 inches and 1.2 inches of precipitation, respectively, in June—well below the 30-year average (see pages 77 and 79). Despite this stress, minimal pollination issues were reported, thanks to timely rainfall in July. Harvest progressed rapidly due to dry conditions in September and October. Purdue's research farms in West Lafayette and Wanatah, IN, recorded only 0.2 inches and 1.8 inches of rainfall in October. As a result, by October 21, 2024, the USDA reported that Indiana's corn grain harvest was 61% complete—16 percentage points ahead of the five-year average.

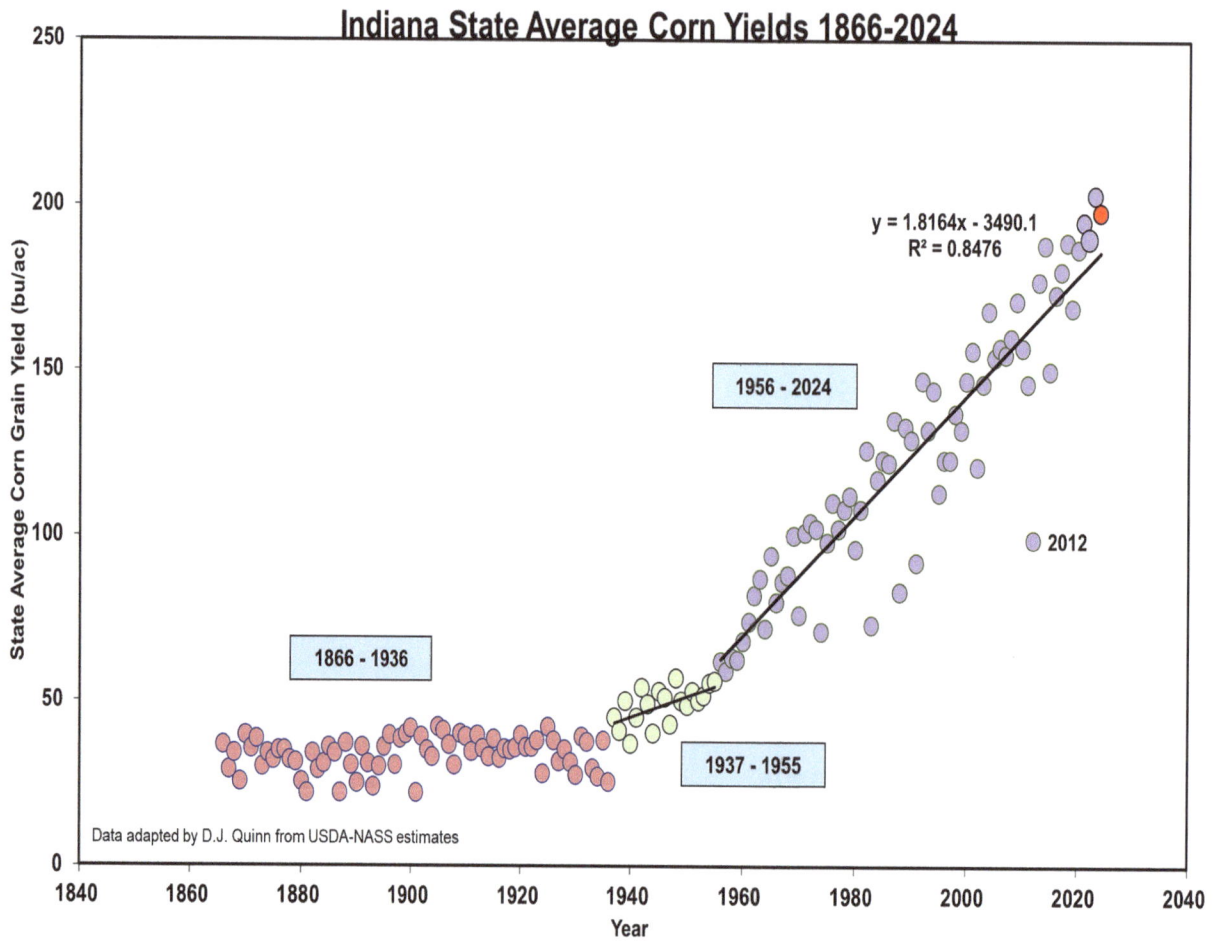

FIGURE 1. Historic state average corn grain yield trends for Indiana (1866–2024). The red point illustrates the 2024 state yield average.

RANDOMIZATION, REPLICATION, AND STATISTICS

MAKING SENSE OF APPLIED FIELD RESEARCH

Field research trials are an important part of understanding how specific agronomic practices can improve farm productivity. Universities such as Purdue use both research station and on-farm research trials across the state to help drive our recommendations and provide management information for Indiana farmers. However, some of our research practices and conclusions may differ from various private-sector research trials and potentially what you may see on your own farm. For example, you may ask: "Why did they set up the research trial that way?," "What are those letters next to the yield values they are presenting?," and "Why does it seem the university never sees any yield responses from various products?" Therefore, it is important to understand how we approach field research trials, the steps we take to determine our conclusions, and how understanding these approaches can help you better understand and test practices more accurately on your own farm.

Two of the first questions I often ask people when discussing research results are: (1) Do you have a yield monitor in your combine? (2) When traveling across the field during harvest, do those yield values stay the same? The answer I receive 100% of the time is no (if yes, you may need to consider a new monitor), and this is largely due to the variability throughout the field caused by soil type differences, elevation differences, and so on. Therefore, when setting up field research trials we often designate a treatment (e.g., new product) and compare that to a nontreated control (e.g., business as usual). And two of the most important questions we ask after harvest are: (1) Was the yield difference observed truly caused by the product we applied? (2) Or was the yield difference only due to the treated areas being in a more productive part of the field? For example, in Figure 2, if I split a field in half and apply my treatment on one half of the field but don't apply my treatment on the other half of the field, I may find a yield difference of 15 bushels per acre and think to myself, "I should apply this product on all of my acres." However, when you look closer, it is easy to see that the treated area of the field encompassed a larger portion of one soil type, whereas the nontreated area encompassed a larger portion of another soil type. Therefore, it is difficult to determine whether the yield response was due to the product applied or due to the treated area being in a more productive area of the field.

In our university research trials, we approach testing a treatment within a field using randomization, replication (repetition of an experiment in similar conditions), and statistics (Figure 3 and Table 1). For example, compare Figure 2 and Figure 3. Figure 3 highlights how we typically set up one of our research trials using replication and randomization of the treated and nontreated passes to account for field differences. Each of these practices helps us improve the reliability of our conclusions, account for random error (e.g., field variability), and determine the true causes of yield differences observed. Furthermore, it is also important for us to perform these research trials across multiple locations and multiple years to determine how treatment

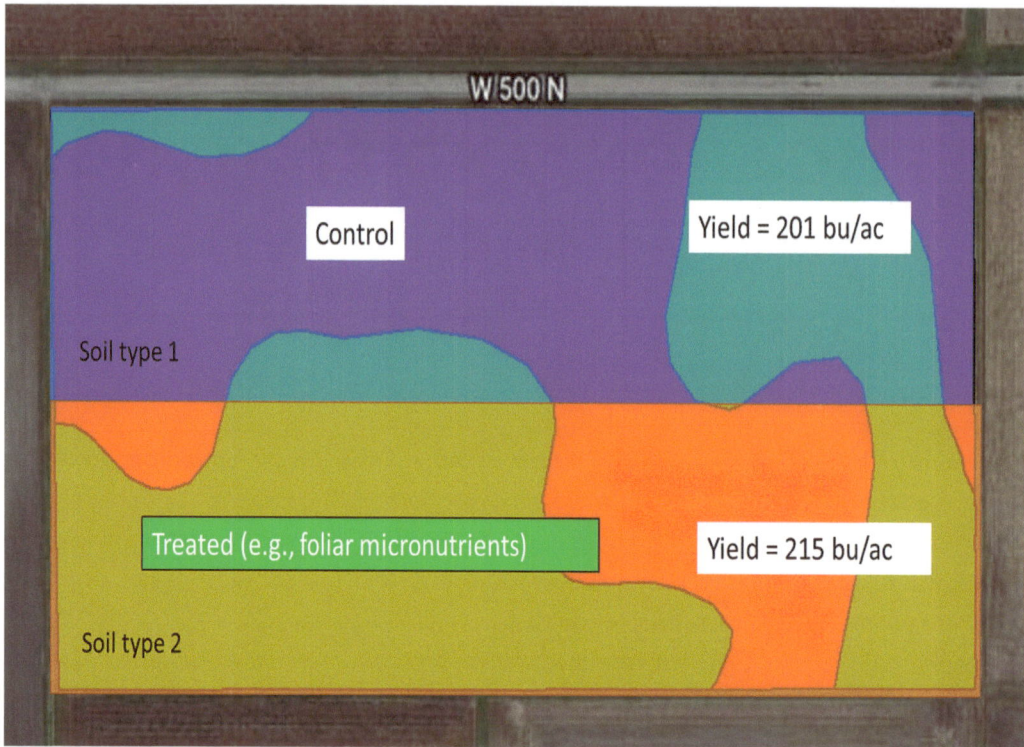

FIGURE 2. Example of a split-field comparison between a nontreated control and a designated treatment.

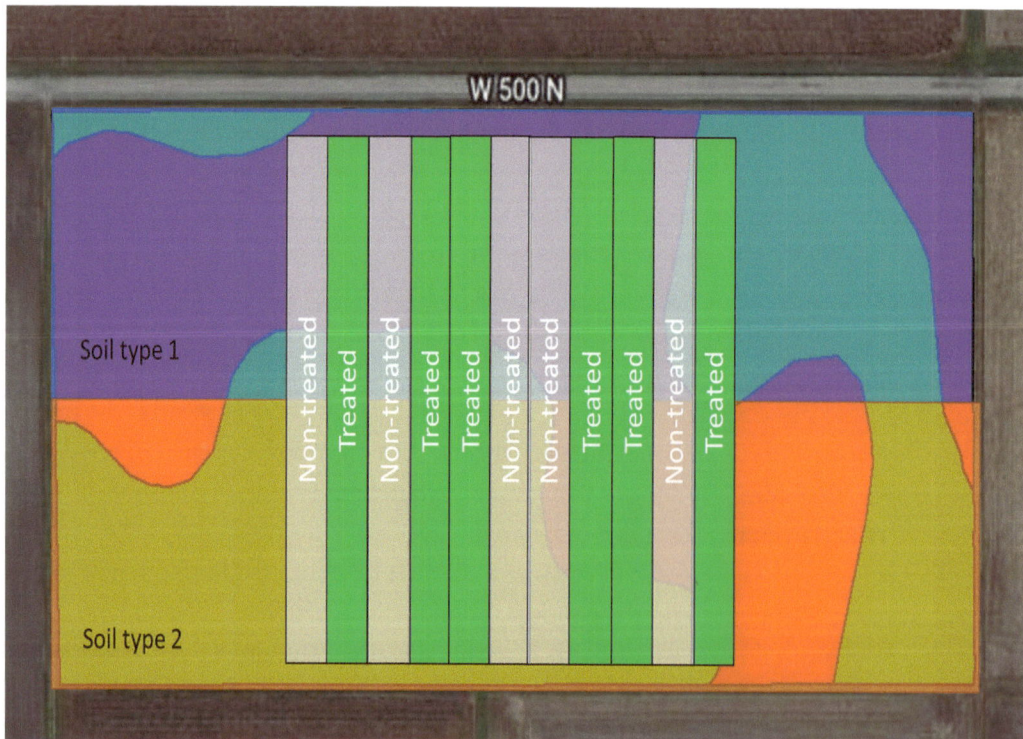

FIGURE 3. Example of a replicated and randomized field research trial comparison between a nontreated control and a designated treatment.

TABLE 1. *Corn grain yield comparisons between the nontreated control and an imposed treatment following a randomized and replicated field research trial.*

TREATMENT	YIELD (BU/AC)
Nontreated	204 a*
Treated	208 a

* Average yield values that contain the same corresponding letters are not statistically different ($P > 0.1$) from each other.

responses may differ in different fields and different environments. We also use statistical models to help determine our conclusions (Table 1). Using statistics helps us determine whether the differences we detect are due to random error or due to the treatment we tested. For example, if you have ever seen university data presented (or the data presented in this report), you have probably seen data presented similar to Table 1. At first glance, after we randomized and replicated our treatments (Figure 3), the treated areas seem to have increased corn yield by 4 bushels per acre (Table 1). However, our conclusions suggested no yield differences were observed. Therefore, through the research steps we implemented, we determined that the yield difference was due to random error (e.g., field variability) and not due to the product or management practice tested. The letters next to the yield values help us highlight where statistical (yield differences due to treatments) differences were observed.

In conclusion, when testing a new product or practice on your own farm, it is important to think about how to design and set up a trial to accurately test the new product or practice. Just because you observe a yield difference doesn't always mean the new product or practice you tested is the reason for this difference. At Purdue, it is our goal to accurately assess new products and practices to determine whether or not these are truly the reason behind observed yield differences. In addition, as you sit in on various meetings and presentations, and examine research results, ask yourself: How did they design and set up this research trial? Did they use randomization, replication, and statistics, and if not, are the yield differences being discussed truly due to the product applied? Over how many different environments and years was this product tested? Understanding and asking these questions can help you determine the best products and management practices to implement and improve your operation.

AGRONOMY CENTER FOR RESEARCH AND EDUCATION (ACRE)

CORN RESPONSE TO NITROGEN FERTILIZER APPLICATION METHODS AND SLOW-RELEASE NITROGEN FERTILIZER PRODUCTS (ACRE)

Daniel Quinn: Department of Agronomy, Purdue University

Rachel Stevens: Purdue Agronomy Center for Research and Education (ACRE)

Aaron Kult: Purdue Agronomy Center for Research and Education (ACRE)

Evan Bossung: Purdue Agronomy Center for Research and Education (ACRE)

Study Location: Purdue Agronomy Center for Research and Education, West Lafayette, IN

Soil Type: Raub-Brenton silt loam–silty clay loam (0–1% slope), Drummer silty clay loam (0–2% slope)

Planting Date: May 03, 2024 | **Harvest Date:** Oct 07, 2024

Corn Hybrid: Pioneer P1108Q | **Corn Seeding Rate:** 32,000 seeds/ac

Corn Nitrogen (N) Fertilizer Rate and Source: See research treatments.

Previous Crop: Soybean | **Tillage:** Spring Vertical Tillage

Study Replications: 5

RESEARCH TRIAL OVERVIEW:

This field research trial was conducted at the Purdue Agronomy Center for Research and Education (ACRE) in Tippecanoe County, IN. The trial evaluated corn yield response to the slow-release liquid nitrogen fertilizer product NitroGain from Arclin added to a 2x2 starter, in-furrow starter, and coulter-injected sidedress. The trial compared different NitroGain application methods to nitrogen sources 28% UAN and ESN. The trial was designed as a randomized complete block design with six treatments and five replications. Plots measure 15 feet wide (six 30-inch rows) by 40 feet long and the center two rows were harvested using a Wintersteiger Plot Combine combined with a HarvestMaster weigh system and adjusted to 15.5% moisture for yield analysis.

RESEARCH TREATMENTS:

1. Nontreated Control (0 lbs N/ac)
2. 2x2 (40 lbs N/ac as 28% UAN) + V5 sidedress (140 lbs N/ac as 28% UAN)
3. 2x2 (40 lbs N/ac as 28% UAN) + V5 sidedress (140 lbs N/ac as NitroGain)
4. 2x2 (40 lbs N/ac as NitroGain) + V5 sidedress (140 lbs N/ac as 28% UAN)
5. In-Furrow (15 lbs N/ac as NitroGain + V5 sidedress (165 lbs N/ac as 28% UAN)
6. Broadcast Preplant (1 week before planting) Dry Fertilizer (180 lbs N/ac as ESN)

RESULTS:

TABLE 2. *Total nitrogen (N) fertilizer rate applied (lbs N/ac), mean grain moisture (%), and mean grain yield (bu/ac) values in response to applied treatments. West Lafayette, IN 2024.*

TREATMENT DESCRIPTION	NITROGEN FERTILIZER RATE	GRAIN MOISTURE	GRAIN YIELD
	---- lbs N/ac ----	---- % ----	---- bu/ac ---
2x2 + V5 SD (UAN)	180	19.5 a*	268.2 a
2x2 + V5 SD (NitroGain)	180	19.6 a	269.5 a
2x2 + V5 SD (NitroGain + UAN)	180	19.8 a	272.8 a
IF + V5 SD (NitroGain + UAN)	180	19.7 a	275.6 a
Preplant (ESN)	180	19.5 a	266.9 a
Pr >F	-	-	0.213
Nontreated Control[†]	0	18.1	151.7

* Mean values that contain dissimilar letters and are in the same column are determined significantly different from each other ($P < 0.1$).
† Nontreated control was excluded from the statistical analysis.

SUMMARY (TAKE-HOME POINTS):

• No statistical differences were observed in mean grain moisture and mean grain yield values between the different nitrogen application methods and sources used in this research trial in 2024 (Table 2).

EVALUATION OF CORN YIELD AND NITROGEN UPTAKE TO A CHLORIDE-FREE AMMONIUM NITRATE-BASED FERTILIZER (28-0-5-6S) (ACRE)

Daniel Quinn: Department of Agronomy, Purdue University

Rachel Stevens: Purdue Agronomy Center for Research and Education (ACRE)

Aaron Kult: Purdue Agronomy Center for Research and Education (ACRE)

Evan Bossung: Purdue Agronomy Center for Research and Education (ACRE)

Study Location: Purdue Agronomy Center for Research and Education, West Lafayette, IN

Soil Type: Raub-Brenton silt loam–silty clay loam (0–1% slope), Drummer silty clay loam (0–2% slope)

Planting Date: May 03, 2024 | **Harvest Date:** Oct 09, 2024

Corn Hybrid: Pioneer P1108Q | **Corn Seeding Rate:** 32,000 seeds/ac

Corn Nitrogen (N) Fertilizer Rate and Source: Various at-plant nitrogen rates with Urea, AMS, and NKS 28

Previous Crop: Soybean | **Tillage:** Spring Vertical Tillage

Study Replications: 5

RESEARCH TRIAL OVERVIEW:

This field research trial was conducted at the Purdue Agronomy Center for Research and Education (ACRE) in Tippecanoe County, IN. The trial examined corn yield response and nitrogen uptake of a new chloride-free, ammonium nitrate–based fertilizer (28-0-5-6S) compared to standard granular fertilizer products. Anvol urease inhibitor was applied at a rate of 1.5 qt/ton of urea on specific treatments below. The trial was designed as a randomized complete block design with 13 treatments and five replications. Plots measure 15 feet wide (six 30-inch rows) by 40 feet long and the center two rows were harvested using a Wintersteiger Plot Combine equipped with a HarvestMaster weigh system and adjusted to 15.5% moisture for yield analysis.

RESEARCH TREATMENTS:

1. Nontreated (No nitrogen fertilizer)
2. Broadcast Preplant (1 week before planting) Urea (120 lbs N/ac)
3. Broadcast Preplant (1 week before planting) Urea (160 lbs N/ac)
4. Broadcast Preplant (1 week before planting) Urea (200 lbs N/ac)
5. Broadcast Preplant (1 week before planting) Urea + Anvol (120 lbs N/ac)
6. Broadcast Preplant (1 week before planting) Urea + Anvol (160 lbs N/ac)
7. Broadcast Preplant (1 week before planting) Urea + Anvol (200 lbs N/ac)
8. Broadcast Preplant Urea + Anvol (97.5 lbs N/ac) + AMS (22.5 lbs N/ac)
9. Broadcast Preplant Urea + Anvol (130 lbs N/ac) + AMS (30 lbs N/ac)
10. Broadcast Preplant Urea + Anvol (162.5 lbs N/ac) + AMS (37.5 lbs N/ac)
11. Broadcast Preplant NKS 28 (28-0-5-6S; 120 lbs N/ac)
12. Broadcast Preplant NKS 28 (28-0-5-6S; 160 lbs N/ac)
13. Broadcast Preplant NKS 28 (28-0-5-6S) 200 lbs N/ac)

RESULTS:

TABLE 3. *Total nitrogen (N) and sulfur (S) fertilizer rate applied (lbs N/ac and lbs S/ac), mean grain moisture (%), and mean grain yield (bu/ac) values in response to applied treatments. West Lafayette, IN 2024.*

TREATMENT DESCRIPTION	N FERTILIZER RATE	S FERTILIZER RATE	GRAIN MOISTURE	GRAIN YIELD
	-- lbs N/ac --	-- lbs S/ac --	---- % ----	---- bu/ac ---
Preplant Urea	120	0	16.5 c*	238.4 f
Preplant Urea	160	0	16.6 c	258.4 de
Preplant Urea	200	0	16.9 abc	263.6 cde
Preplant Urea + Anvol	120	0	16.8 abc	251.2 ef
Preplant Urea + Anvol	160	0	17.3 a	278.6 abc
Preplant Urea + Anvol	200	0	17.3 ab	269.4 bcd
Preplant Urea + Anvol + AMS	120	26	16.8 abc	268.9 bcd
Preplant Urea + Anvol + AMS	160	34	16.7 bc	276.0 abc
Preplant Urea + Anvol + AMS	200	43	17.0 abc	288.6 a
NKS 28	120	21	16.5 c	265.5 cde
NKS 28	160	29	17.1 abc	284.1 ab
NKS 28	200	36	17.3 a	289.2 a
Pr > F	–	–	0.023	0.001
Nontreated Control[†]	0	0	16.5	162.1

* Mean values that contain dissimilar letters and are in the same column are determined significantly different from each other ($P < 0.1$).
† Nontreated control was excluded from the statistical analysis.

TABLE 4. *Mean corn grain moisture (%) and grain yield (bu/ac) values in response to fertilizer products. Data combined across N and S rates applied. West Lafayette, IN 2024.*

TREATMENT DESCRIPTION	GRAIN MOISTURE	GRAIN YIELD
	---- % ----	---- bu/ac ---
Preplant Urea	16.7 a*	253.4 c
Preplant Urea + Anvol	17.1 a	266.4 b
Preplant Urea + Anvol + AMS	16.8 a	277.9 a
NKS 28	16.9 ab	279.6 a
Pr > F	0.291	0.004

* Mean values that contain dissimilar letters and are in the same column are determined significantly different from each other ($P < 0.1$).

SUMMARY (TAKE-HOME POINTS):

- Mean corn grain yield and grain moisture were statistically increased as N fertilizer rate increased (Table 3). Across all treatments, no additional grain yield was gained by increasing the N fertilizer rate from 160 lbs N/ac to 200 lbs N/ac. Also, keep in mind that the S fertilizer rate also increased as N fertilizer rate increased for the treatments containing AMS and NKS 28.
- Across all N and S fertilizer rates applied, yield was increased by the inclusion of Anvol by 13 bu/ac, Anvol + AMS by 24 bu/ac, and NKS 28 by 26 bu/ac in comparison to only a preplant broadcast application of urea (Table 4). Results suggest a high probability of ammonia volatilization due to urea application method and a significant response to S at this location.

CORN YIELD RESPONSE TO IN-FURROW STARTER AND IN-SEASON FOLIAR APPLICATION PRODUCTS AT VARIOUS APPLICATION TIMINGS (ACRE)

Daniel Quinn: Department of Agronomy, Purdue University
Rachel Stevens: Purdue Agronomy Center for Research and Education (ACRE)
Aaron Kult: Purdue Agronomy Center for Research and Education (ACRE)
Evan Bossung: Purdue Agronomy Center for Research and Education (ACRE)

Study Location: Purdue Agronomy Center for Research and Education, West Lafayette, IN

Soil Type: Raub-Brenton silt loam–silty clay loam (0–1% slope)

Planting Date: May 08, 2024 | **Harvest Date:** Oct 09, 2024

Corn Hybrid: Pioneer P1108Q | **Corn Seeding Rate:** 32,000 seeds/ac

Corn Nitrogen (N) Fertilizer Rate, Source, and Timing: No starter (2x2 or in-furrow) fertilizer N was applied. Remaining N fertilizer was applied at 180 lbs N/ac as a coulter-injected sidedress application at the V4 growth stage as 28% UAN (same across all treatments).

Previous Crop: Soybean | **Tillage:** Spring Vertical Tillage

Study Replications: 5

RESEARCH TRIAL OVERVIEW:

This field research trial was conducted at the Purdue Agronomy Center for Research and Education (ACRE) in Tippecanoe County, IN. The trial examined corn yield response to Valent BioSciences Symvado SC, Transit Foliar, and Delaro Complete fungicide across various application methods. All in-season foliar spray combinations were tank-mixed and applied simultaneously. The trial was designed as a randomized complete block design with six treatments and five replications. Plots measure 15 feet wide (six 30-inch rows) by 40 feet long and the center two rows were harvested using a Wintersteiger Plot Combine equipped with a HarvestMaster weigh system and adjusted to 15.5% moisture for yield analysis.

RESEARCH TREATMENTS:

1. Nontreated Control (No in-furrow or in-season foliar treatments)
2. Transit Foliar (10 oz/ac) at V3/V5 applied at 15 gal/ac
3. Symvado SC in-furrow (2 oz/ac) applied at 5 gal/ac
4. Delaro Complete (R1 growth stage, 10 oz/ac) at 20 gal/ac
5. Delaro Complete (R1 growth stage, 10 oz/ac) + Transit Foliar (10 oz/ac) at 20 gal/ac
6. Delaro Complete (R1 growth stage, 10 oz/ac) + Transit Foliar (15 oz/ac) at 20 gal/ac

RESULTS:

TABLE 5. *Mean corn grain moisture (%) and grain yield (bu/ac) values in response to applied in-furrow and foliar application treatments. West Lafayette, IN 2024.*

TREATMENT DESCRIPTION	GRAIN MOISTURE	GRAIN YIELD
	---- % ----	---- bu/ac ---
Nontreated Control	20.3 a*	274.0 a
Transit Foliar (V3-V5)	20.1 a	280.9 a
Symvado SC (In-furrow)	19.7 a	279.0 a
Delaro Complete (10 oz/ac at R1)	20.1 a	278.6 a
Delaro Complete + Transit Foliar (10 oz/ac at R1)	20.1 a	280.6 a
Delaro Complete + Transit Foliar (15 oz/ac at R1)	20.3 a	283.2 a
Pr > F	0.610	0.284

* Mean values that contain dissimilar letters and are in the same column are determined significantly different from each other ($P < 0.1$).

SUMMARY (TAKE-HOME POINTS):

- No statistical yield differences were observed across treatments in this research trial (Table 5). Minimal foliar disease pressure was also observed in this research trial, which likely impacted the lack of R1 fungicide response.

CORN RESPONSE TO AT-PLANT IN-FURROW AND SEED-APPLIED FERTILIZER PRODUCTS (ACRE)

Daniel Quinn: Department of Agronomy, Purdue University
Rachel Stevens: Purdue Agronomy Center for Research and Education (ACRE)
Aaron Kult: Purdue Agronomy Center for Research and Education (ACRE)
Evan Bossung: Purdue Agronomy Center for Research and Education (ACRE)

Study Location: Purdue Agronomy Center for Research and Education, West Lafayette, IN
Soil Type: Raub-Brenton silt loam–silty clay loam (0–1% slope), Drummer silty clay loam (0–2% slope)
Planting Date: May 08, 2024 | **Harvest Date:** Oct 09, 2024
Corn Hybrid: Pioneer P1108Q | **Corn Seeding Rate:** 32,000 seeds/ac
Corn Nitrogen (N) Fertilizer Rate and Source: No additional N fertilizer was applied via a starter application (2x2 or in-furrow) beyond the research treatments below. Remaining N fertilizer was applied at a 180 lbs N/ac as a coulter-injected sidedress application at the V5 growth stage as 28% UAN (same across all treatments).
Previous Crop: Soybean | **Tillage:** Spring Vertical Tillage
Study Replications: 5

RESEARCH TRIAL OVERVIEW:

This field research trial was conducted at the Purdue Agronomy Center for Research and Education (ACRE) in Tippecanoe County, IN. The trial examined corn grain moisture and yield differences of in-furrow and seed-applied nutritional products from Brandt. The trial was designed as a randomized complete block design with five replications. All plots measured 15 feet wide (six 30-inch corn rows) x 40 feet long and the center two rows of each plot were harvested with a Wintersteiger small-plot combine equipped with a Harvest-Master weigh system and adjusted to 15.5% moisture for yield analysis.

RESEARCH TREATMENTS:

1. Nontreated control (no in-furrow products applied)
2. Ammonium Polyphosphate (APP, 10-34-0, in-furrow at 5 gal/ac)
3. EnzUP Phosphorus DS only (in-furrow at 5 lbs/ac, dry soluble product)
4. APP (in-furrow at 2.5 gal/ac) + EnzUP Phosphorus DS (in-furrow at 2.5 lbs/ac). Rates were reduced for both products to apply same total P as treatments 2 and 3.
5. APP (in-furrow at 5 gal/ac) + EnzUP Zinc (in-furrow at 2 qt/ac)
6. APP (in-furrow at 5 gal/ac) + EnzUP Seedflow Zinc (seed-applied at 2 oz/80,000 seed unit)
7. EnzUP Phosphorus DS (in-furrow at 2.5 lbs/ac)

RESULTS:

TABLE 6. *Corn grain moisture (%) and yield (bu/ac) differences observed from applied to in-furrow and seed-applied products. West Lafayette, IN 2024.*

AT-PLANT TREATMENT	GRAIN MOISTURE	GRAIN YIELD
	---- % ----	---- bu/ac ----
Nontreated	23.8 a*	288.5 ab
Ammonium Polyphosphate (10-34-0, 5 gal/ac)	23.8 a	281.3 c
EnzUP P DS (5 lbs/ac)	23.6 a	288.6 ab
APP (2.5 gal/ac) + EnzUP P DS (2.5 lb/ac)	23.4 a	291.9 a
APP (5 gal/ac) + EnzUP Zn (1 qt/ac)	23.1 a	292.1 a
APP (5 gal/ac) + EnzUP SeedFlow Zn (2 oz/unit)	23.7 a	288.8 ab
EnzUP P DS (2.5 lb/ac)	23.8 a	285.6 bc
Pr > F	0.559	*0.016*

* Mean values that do not contain the same letter are determined statistically different from each other ($P < 0.1$).

SUMMARY (TAKE-HOME POINTS):

- No statistical yield increases were observed across treatments in this research trial in comparison to the nontreated control (Table 6). Results did observe a slight yield decrease with an application of Ammonium Polyphosphate (10-34-0) when applied alone and in-furrow at a rate of 5 gal/ac. However, current available data failed to provide a reason for this response.

CORN RESPONSE TO INSTINCT NXTGEN® ACROSS N FERTILIZER SOURCES, RATES, AND TIMINGS (ACRE)

Emely Gramajo: Department of Agronomy, Purdue University

Daniel Quinn: Department of Agronomy, Purdue University

Rachel Stevens: Purdue Agronomy Center for Research and Education (ACRE)

Aaron Kult: Purdue Agronomy Center for Research and Education (ACRE)

Evan Bossung: Purdue Agronomy Center for Research and Education (ACRE)

Study Location: West Lafayette, IN.

Soil Type: Raub-Brenton complex, Fine-Silty (0–1% slope)

Planting Date: May 04, 2024 | **Harvest Date:** October 08, 2024

Corn Hybrid: Pioneer P1108Q | **Corn Seeding Rate:** 32,000 seeds/ac

Corn Nitrogen (N) Fertilizer Rate and Source: 0, 120, and 180 lbs N/ac, UAN (28-0-0; sidedress only) and Urea (46-0-0; preplant applications only). No starter N fertilizer was included in this trial.

Previous Crop: Soybean | **Tillage:** Conventional

Study Replications: 5

RESEARCH TRIAL OVERVIEW:

This research trial was established at the Agronomy Center for Research and Education (ACRE) in Tippecanoe County, IN. This research trial examined corn grain yield differences in response to various N fertilizer sources, rates, and timings with and without nitrification inhibitor Instinct NXTGEN from Corteva Agriscience. The plot was designed as a randomized complete block design with 13 treatments and five replications. All plots measured 15 feet wide (six 30-inch cornrows) x 40 feet long and the center two rows of each plot were harvested using a Wintersteiger Plot Combine equipped with a HarvestMaster weigh system and adjusted to 15.5% moisture for yield analysis.

RESEARCH TREATMENTS:

1. Nontreated control (No N fertilizer applied)
2. 120 lbs of N/ac applied as UAN (28-0-0) at V3
3. 180 lbs of N/ac applied as UAN (28-0-0) at V3
4. 120 lbs of N/ac applied as UAN (28-0-0) at V3 with Instinct NXTGEN at 24 oz/ac
5. 180 lbs of N/ac applied as UAN (28-0-0) at V3 with Instinct NXTGEN at 24 oz/ac
6. 120 lbs of N/ac applied as 60 lbs N/ac Urea preplant and 60 lbs N/ac UAN (28-0-0) at V3
7. 180 lbs of N/ac applied as 90 lbs N/ac Urea preplant and 90 lbs N/ac UAN (28-0-0) with 12 oz/ac of Instinct NXTGEN at V5
8. 120 lbs of N/ac applied in as 60 lbs N/ac Urea preplant with 12 oz/ac of Instinct NXTGEN and 60 lb N/ac UAN (28-0-0) with 12 oz/ac of Instinct NXTGEN at V5
9. 180 lbs of N/ac applied as 90 lbs N/ac Urea preplant with 12 oz/ac of Instinct NXTGEN and 90 lbs N/ac UAN (28-0-0) at V5
10. 120 lbs of N/ac applied as Urea preplant
11. 180 lbs of N/ac applied as Urea preplant

12. 120 lbs of N/ac applied as Urea preplant with Instinct NXTGEN at 24 oz/ac
13. 180 lbs of N/ac applied as Urea preplant with Instinct NXTGEN at 24 oz/ac

RESULTS:

TABLE 7. *Mean corn grain moisture (%) and yield (bu/ac) in response to the interaction between total N fertilizer rate and the inclusion of Instinct NXTGEN. West Lafayette, IN 2024.*

NITROGEN FERTILIZER RATE	INSTINCT NXTGEN	GRAIN MOISTURE	GRAIN YIELD
---lbs/ac---		---%---	---bu/ac---
120	No	16.8 a*	236.3 b
120	Yes	16.7 a	241.0 b
180	No	17.3 a	253.3 a
180	Yes	17.3 a	256.8 a
Nontreated Control (0 lbs N/ac)[†]	-	-	139.2

*Mean values that do not contain the same corresponding letter are determined statistically different ($P < 0.1$). Columns with mean values that do not contain any letters are determined as no statistical differences between treatments.
† Nontreated control was not included in the analysis.

TABLE 8. *Mean corn grain moisture (%) and yield (bu/ac) averages examining the interaction between N fertilizer timing and Instinct NXTGEN. West Lafayette, IN 2024.*

NITROGEN FERTILIZER TIMING	INSTINCT NXTGEN	GRAIN MOISTURE	GRAIN YIELD
---lbs/ac---		---%---	---bu/ac---
Preplant	No	16.7 bc*	223.2 b
Preplant	Yes	16.6 c	230.1 b
V3 Sidedress	No	17.6 a	259.2 a
V3 Sidedress	Yes	17.3 ab	259.9 a
Split Application (Preplant + V6)	No	17.0 abc	252.0 a
Split Application (Preplant + V6)	Yes	17.1 abc	256.9 a
Nontreated Control (0 lbs N/ac)	-	-	139.2

*Mean values that do not contain the same corresponding letter are determined statistically different ($P < 0.1$). Columns with mean values that do not contain any letters are determined as no statistical differences between treatments.
† Nontreated control was not included in the analysis.

SUMMARY (TAKE-HOME POINTS):

- The inclusion of Instinct NXTGEN did not increase corn grain yield across both N fertilizer rates (120 and 180 lbs N/acre) and N fertilizer sources (UAN and urea) in this research trial (Table 7). The inclusion of Instinct NXTGEN did not increase corn grain yield across the different nitrogen fertilizer application methods in this research trial (Table 8). Minimal weather events that would cause N losses (leaching, denitrification) were observed at this location in 2024, which likely drove the overall lack of a response.
- Trial results did observe a significant yield response when comparing N application method only (Table 8). Both the in-season sidedress (V3 only) and the split application (preplant + V6) increased corn grain yield by an average of 33 bu/ac and 28 bu/ac, respectively.

CORN RESPONSE TO IN-FURROW AND FOLIAR COMBINATIONS OF C-CAT (ACRE)

Daniel Quinn: Department of Agronomy, Purdue University
Rachel Stevens: Purdue Agronomy Center for Research and Education (ACRE)
Aaron Kult: Purdue Agronomy Center for Research and Education (ACRE)
Evan Bossung: Purdue Agronomy Center for Research and Education (ACRE)

Study Location: Purdue Agronomy Center for Research and Education, West Lafayette, IN
Soil Type: Raub-Brenton silt loam–silty clay loam (0–1% slope), Drummer silty clay loam (0–2% slope)
Planting Date: May 08, 2024 | **Harvest Date:** Oct 09, 2024
Corn Hybrid: Pioneer P1108Q | **Corn Seeding Rate:** 32,000 seeds/ac
Corn Nitrogen (N) Fertilizer Rate and Source: Nitrogen fertilizer was applied at a rate of 180 lbs N/ac and as a split application with 40 lbs N/ac at planting (2x2 starter) + V4 coulter-injected sidedress (140 lbs N/ac) across all treatments.
Previous Crop: Soybean | **Tillage:** Spring Vertical Tillage
Study Replications: 5

RESEARCH TRIAL OVERVIEW:

This field research trial was established at the Purdue Agronomy Center for Research and Education (ACRE) in Tippecanoe County, IN. The trial examined corn yield response to C-CAT from TEVA applied at different rates and application methods (in-furrow, foliar at R1, and in-furrow followed by foliar at R1). The trial was designed as a randomized complete block design with seven treatments and five replications. Plots measure 15 feet wide (six 30-inch rows) by 40 feet long and the center two rows were harvested using a Wintersteiger Plot Combine equipped with a HarvestMaster weigh system and adjusted to 15.5% moisture for yield analysis.

RESEARCH TREATMENTS:

1. Nontreated Control
2. C-CAT (In-furrow) 1 gal/ac, application rate at 5 gal/ac
3. C-CAT (In-furrow) 2 gal/ac, application rate at 5 gal/ac
4. C-CAT (In-furrow) 4 gal/ac, application rate at 5 gal/ac
5. C-CAT (In-furrow) 1 gal/ac, application rate at 5 gal/ac + C-CAT (foliar at R1) at 1 gal/ac, application rate at 20 gal/ac
6. C-CAT (In-furrow) 2 gal/ac, application rate at 5 gal/ac + C-CAT (foliar at R1) at 1 gal/ac, application rate at 20 gal/ac
7. C-CAT (foliar at R1) at 3 gal/A

RESULTS:

TABLE 9. *Mean corn grain moisture (%), yield (bu/ac), and plant stand (plants/ac) in response to in-furrow and R1 foliar treatments. West Lafayette, IN 2024.*

TREATMENT NO.	TREATMENT DESCRIPTION	PRODUCT RATE	GRAIN MOISTURE	GRAIN YIELD	PLANT STAND
			-- % --	-- bu/ac --	-- plants/ac --
1	Nontreated Control		19.4 a*	268.5 a	30100 d
2	C-CAT (In-furrow)	1 gal/ac	19.3 a	271.0 a	31102 ab
3	C-CAT (In-furrow)	2 gal/ac	19.2 a	272.4 a	31537 cd
4	C-CAT (In-furrow)	4 gal/ac	19.6 a	272.2 a	30579 bcd
5	C-CAT (In-furrow)	1 gal/ac	19.3 a	266.6 a	31058 ab
	C-CAT (R1 Foliar)	1 gal/ac			
6	C-CAT (In-furrow)	2 gal/ac	19.3 a	270.8 a	31276 ab
	C-CAT (R1 Foliar)	1 gal/ac			
7	C-CAT (R1 Foliar)	3 gal/ac	19.2 a	270.2 a	30318 cd
Pr > F			0.633	0.938	0.023

* Mean values that contain dissimilar letters and are in the same column are determined significantly different from each other (*P* < 0.1).

SUMMARY (TAKE-HOME POINTS):

- Corn plant stand was improved with the application of C-CAT applied in-furrow, primarily at the rate of 1 gal/ac (Table 9). However, despite the improvements in final plant stand, no yield differences were observed across all applied treatments.

COMPARISON OF SHORT- AND TALL-STATURE CORN HYBRIDS TO NITROGEN FERTILIZER AND SEEDING RATES (ACRE)

Erick Oliva: Department of Agronomy, Purdue University

Rachel Stevens: Purdue Agronomy Center for Research and Education (ACRE)

Aaron Kult: Purdue Agronomy Center for Research and Education (ACRE)

Evan Bossung: Purdue Agronomy Center for Research and Education (ACRE)

Daniel Quinn: Department of Agronomy, Purdue University

Study Location: Purdue Agronomy Center for Research and Education, West Lafayette, IN.

Soil Type: Chalmers silty clay loam (0–2% slope) and Raub-Breton complex (0–1% slope)

Planting Date: May 04, 2024 | **Harvest Date:** October 05, 2024

Corn Hybrid: Preceon PR111-20SSC, Preceon PR112-20SSC, and Dekalb DKC61-41RIB, Dekalb DKC62-70RIB

Corn Seeding Rate: 32,000, 38,000 and 44,000 seeds/ac

Corn Nitrogen (N) Fertilizer Rate and Source: 0, 90, 180 and 270 lbs N/ac coulter-injected as sidedress application of 28% UAN at the V5 growth stage. No starter N was used in this trial.

Previous Crop: Soybean | **Tillage:** Spring Vertical Tillage

Study Replications: 5

RESEARCH TRIAL OVERVIEW:

A field research trial was established at the Purdue Agronomy Center for Research and Education (ACRE) in Tippecanoe County, IN. The trial examined corn yield response to different short- and full-size corn hybrids, nitrogen, and seeding rates. The trial was designed as a split-plot, randomized complete block design with 48 treatments and five replications. Plots measured 15 feet wide (six 30-inch corn rows) by 40 feet long and the center two rows were harvested using a Wintersteiger plot combine equipped with a HarvestMaster weigh system and adjusted to 15.5% moisture for yield analysis.

RESULTS:

TABLE 10. *Short-stature and tall-stature corn grain yield (bu/ac) response to hybrid and seeding rate. West Lafayette, IN 2024.*

TARGETED SEED RATE	HYBRID (TYPE)	GRAIN YIELD
		-- bu/ac --
32,000 seeds/ac	PR111-20SSC (Short)	238 b*
	PR112-20SSC (Short)	230 c
	DKC61-41RIB (Tall)	251 a
	DKC62-70RIB (Tall)	257 a
38,000 seeds/ac	PR111-20SSC (Short)	243 c
	PR112-20SSC (Short)	236 c
	DKC61-41RIB (Tall)	255 b
	DKC62-70RIB (Tall)	262 a
44,000 seeds/ac	PR111-20SSC (Short)	242 b
	PR112-20SSC (Short)	238 b
	DKC61-41RIB (Tall)	254 a
	DKC62-70RIB (Tall)	258 a

* Mean values that do not contain the same corresponding letter and are within the same column and seeding rate are determined statistically different ($P < 0.1$).

TABLE 11. *Short-stature and tall-stature corn grain moisture (%) and yield (bu/ac) in response to hybrid and nitrogen (N) fertilizer rate. West Lafayette, IN 2024.*

HYBRID	NITROGEN RATE	GRAIN MOISTURE	GRAIN YIELD
	--lbs N/ac--	-- % --	-- bu/ac --
PR111-20SSC	0	18.1 c*	151 c
	90	19.3 b	245 b
	180	20.3 a	284 a
	270	20.7 a	283 a
PR112-20SSC	0	16.3 c	144 c
	90	18.0 b	243 b
	180	18.9 a	276 a
	270	19.3 a	276 a
DKC61-41RIB	0	18.7 c	169 c
	90	19.4 b	261 b
	180	19.9 b	293 a
	270	21.0 a	291 a
DKC62-70RIB	0	17.2 b	174 c
	90	17.8 b	264 b
	180	18.5 a	297 a
	270	19.0 a	302 a

* Mean values that do not contain the same corresponding letter and are within the same column and hybrid type are determined statistically different ($P < 0.1$).

SUMMARY (TAKE-HOME POINTS):

- Across all examined treatments (populations and N rates), full-stature hybrids yielded higher than short-stature hybrids by 15–20 bu/ac. Across all examined treatments, both short-stature hybrids had statistically lower final plant stands (data not shown).
- No interactions were observed between hybrid type and N fertilizer rate applied (Table 11). Results suggest all hybrids examined in this trial respond similarly to increases in N fertilizer rate.
- A significant interaction between hybrid type and seeding rate was observed in this research trial (Table 10). Results observed yield differences in short-stature hybrids at the 32,000 seeds/ac rate and the tall-stature hybrids at the 38,000 seeds/ac rate.

CORN RESPONSE TO HYBRID TYPE AND FERTILIZER APPLICATION METHOD (ACRE)

Daniel Quinn: Department of Agronomy, Purdue University

Rachel Stevens: Purdue Agronomy Center for Research and Education (ACRE)

Aaron Kult: Purdue Agronomy Center for Research and Education (ACRE)

Evan Bossung: Purdue Agronomy Center for Research and Education (ACRE)

Study Location: Purdue Agronomy Center for Research and Education, West Lafayette, IN

Soil Type: Raub-Brenton silt loam–silty clay loam (0–1% slope), Drummer silty clay loam (0-2% slope)

Planting Date: May 03, 2024 | **Harvest Date:** Oct 09, 2024

Corn Hybrid: Becks "6184V2P," "6414V2P," "6152D1," "6274VP"

Corn Seeding Rate: 32,000 seeds/ac

Corn Nitrogen (N) Fertilizer Rate and Source: See research treatments below. No starter N fertilizer (2x2 or in-furrow) was applied in this research trial.

Previous Crop: Soybean | **Tillage:** See research treatments below.

Study Replications: 5

RESEARCH TRIAL OVERVIEW:

A field research trial was established at the Purdue Agronomy Center for Research and Education (ACRE) in Tippecanoe County, IN. The research trial examined corn response to four different Becks Hybrids with different classified root architectures (vertical vs. horizontal) with strip-till banded or broadcast-applied Super U (46-0-0) + MESZ (12-40-0-10S-1Zn) at a rate equivalent to 180 lbs N/ac applied on April 22, 2024. The trial was designed with randomized completed block design with eight treatments and five replications. Plot size measured 15 feet wide (six 30-inch corn rows) x 40 feet long and the center two rows were harvested using a Wintersteiger Plot Combine equipped with a HarvestMaster weigh system and adjusted to 15.5% moisture for yield analysis.

RESEARCH TREATMENTS:

1. Becks 6184V2P (horizontal roots) Super U (46-0-0 at 156 lbs N/ac) + MESZ (12-40-0-10S-1Zn at 24 lbs N/ac) preplant broadcast applied and incorporated 2–3" deep
2. Becks 6414V2P (vertical roots) Super U (46-0-0 at 156 lbs N/ac) + MESZ (12-40-0-10S-1Zn at 24 lbs N/ac) preplant broadcast applied and incorporated 2–3" deep
3. Becks 6152D1 (horizontal roots) Super U (46-0-0 at 156 lbs N/ac) + MESZ (12-40-0-10S-1Zn at 24 lbs N/ac) preplant broadcast applied and incorporated 2–3" deep
4. Becks 6274VP (horizontal roots) Super U (46-0-0 at 156 lbs N/ac) + MESZ (12-40-0-10S-1Zn at 24 lbs N/ac) preplant broadcast applied and incorporated 2–3" deep
5. Becks 6184V2P (horizontal roots) Super U (46-0-0 at 156 lbs N/ac) + MESZ (12-40-0-10S-1Zn at 24 lbs/N ac) applied in a strip-till banded application at 3–4" deep

6. Becks 6414V2P (vertical roots) Super U (46-0-0 at 156 lbs N/ac) + MESZ (12-40-0-10S-1Zn at 24 lbs/N ac) applied in a strip-till banded application at 3–4" deep
7. Becks 6152D1 (horizontal roots) Super U (46-0-0 at 156 lbs N/ac) + MESZ (12-40-0-10S-1Zn at 24 lbs/N ac) applied in a strip-till banded application at 3–4" deep
8. Becks 6274VP (horizontal roots) Super U (46-0-0 at 156 lbs N/ac) + MESZ (12-40-0-10S-1Zn at 24 lbs/N ac) applied in a strip-till banded application at 3–4" deep

RESULTS:

TABLE 12. *Corn V5 whole plant biomass, N uptake, final plant population, and yield differences observed between hybrid type and fertilizer application method. West Lafayette, IN 2024.*

CORN HYBRID	FERTILIZER APPLICATION TYPE	V5 WHOLE-PLANT DRY MATTER	V5 WHOLE-PLANT N UPTAKE	PLANT POPULATION	GRAIN YIELD
		--- g/10 plants --	-- lbs N/ac --	--- plants/ac ---	--- bu/ac ---
Becks 6184V2P (Horizontal)	Broadcast	75.6 bc*	43.8 b	31631 c	307.5 a
	Banded	80.7 ab	45.3 ab	32391 ab	284.5 bc
Becks 6414V2P (Vertical)	Broadcast	69.4 c	37.8 cd	31810 bc	287.2 b
	Banded	85.6 a	49.6 a	31765 bc	274.6 cd
Becks 6152D1 (Horizontal)	Broadcast	70.5 c	40.1 bc	32748 a	289.6 b
	Banded	72.1 bc	40.9 bc	32614 a	255.7 e
Becks 6274VP (Horizontal)	Broadcast	58.4 d	32.9 d	32614 a	286.8 b
	Banded	54.9 d	32.1 d	32346 ab	269.5 d

* Mean values within the same column that do not contain the same letter and are within the same column are determined significantly different from each other ($P < 0.1$).

SUMMARY (TAKE-HOME POINTS):

- Across all hybrids examined, the strip-till banded nutrient application yielded lower than the broadcast nutrient application (Table 12).
- However, the hybrid with a classified "vertical" root system exhibited a lower magnitude yield reduction due to a strip-till banded nutrient application and also improved V5 plant nutrient uptake in comparison to hybrids with classified "horizontal" root systems (Table 12), indicating a potential hybrid x nutrient placement interaction that needs to be further explored.
- Research trials will be expanded in 2025 to examine additional treatments and locations.

CORN YIELD RESPONSE TO HYBRID MATURITY AND PLANTING DATE (ACRE)

Daniel Quinn: Department of Agronomy, Purdue University

Rachel Stevens: Purdue Agronomy Center for Research and Education (ACRE)

Aaron Kult: Purdue Agronomy Center for Research and Education (ACRE)

Evan Bossung: Purdue Agronomy Center for Research and Education (ACRE)

Study Location: Purdue Agronomy Center for Research and Education, West Lafayette, IN

Soil Type: Raub-Brenton silt loam–silty clay loam (0–1% slope), Drummer silty clay loam (0–2% slope)

Planting Date: See treatments below | **Harvest Date:** Oct 07, 2024

Corn Hybrid: Pioneer P1108Q and P9608Q | **Corn Seeding Rate:** 32,000 seeds/ac

Corn Nitrogen (N) Fertilizer Rate and Source: Nitrogen fertilizer was applied at a rate of 180 lbs N/ac and as a split application with 40 lbs N/ac at planting (2x2 starter) + V4 coulter-injected sidedress (140 lbs N/ac) across all treatments. The sidedress applications were not applied on the same day across all treatments, but were applied to the specific growth stage of each treatment, which differed due to different planting dates.

Previous Crop: Soybean | **Tillage:** Spring Vertical Tillage

Study Replications: 5

RESEARCH TRIAL OVERVIEW:

A field research trial was conducted at the Purdue Agronomy Center for Research and Education (ACRE) in Tippecanoe County, IN. The research trial examined corn yield response to planting date of two corn hybrids. The trial was designed as a split-plot, four seeding date treatments for a total of eight treatments, two hybrids in the split, and five replications. Plots measure 15 feet wide (six 30-inch rows) by 40 feet long and the center two rows were harvested using a Wintersteiger Plot Combine equipped with a HarvestMaster weigh system and adjusted to 15.5% moisture for yield analysis. Planting dates 1 and 2 were harvested on October 7, planting date 3 was harvested on October 14, and planting date 4 was harvested on October 30, 2024.

RESEARCH TREATMENTS:

1. Pioneer P1108Q planted on April 22
2. Pioneer P9608Q planted on April 22
3. Pioneer P1108Q planted on May 13
4. Pioneer P9608Q planted on May 13
5. Pioneer P1108Q planted on June 10
6. Pioneer P9608Q planted on June 10
7. Pioneer P1108Q planted on June 21
8. Pioneer P9608Q planted on June 21

RESULTS:

TABLE 13. *Mean corn grain moisture (%) and yield (bu/ac) in response to different planting dates. Data was combined across hybrids. West Lafayette, IN 2024.*

CORN PLANTING DATE	GRAIN MOISTURE	GRAIN YIELD
	---- % ----	---- bu/ac ---
Planting Date 1 (April 22)	14.9 c*	235.9 b
Planting Date 2 (May 13)	16.9 b	256.1 a
Planting Date 3 (June 10)	19.8 a	162.1 c
Planting Date 4 (June 21)	20.7 a	117.4 d
$Pr > F$	0.001	0.001

* Mean values within the same column that do not contain the same letter and are within the same column are determined significantly different from each other ($P < 0.1$).

TABLE 14. *Mean corn grain moisture (%) and yield (bu/ac) in response to different hybrid types. Data was combined across planting dates. West Lafayette, IN 2024.*

CORN HYBRID TYPE	GRAIN MOISTURE	GRAIN YIELD
	---- % ----	---- bu/ac ---
P1108Q (110-d maturity)	20.1 a*	223.9 a
P9608Q (96-d maturity)	15.9 b	178.7 b
$Pr > F$	0.001	0.001

* Mean values within the same column that do not contain the same letter and are within the same column are determined significantly different from each other ($P < 0.1$).

SUMMARY (TAKE-HOME POINTS):

- No interaction between hybrid type and planting date was observed in this study. Therefore, only main effect results were presented (Tables 13 and 14).
- The second planting date (May 13) produced the highest grain yield in comparison to all other planting date treatments (Table 13).
- The hybrid with the longer relative maturity date (P1108Q) yielded ~45 bu/ac higher than the shorter relative maturity date hybrid (P9608Q) across all planting dates examined (Table 14).

CORN RESPONSE TO INTENSIVE MANAGEMENT WITH IRRIGATION AND FERTIGATION (ACRE)

Jose Vaca: Department of Agronomy, Purdue University

Laura Bowling: Department of Agronomy, Purdue University

Shaun Casteel: Department of Agronomy, Purdue University

Daniel Quinn: Department of Agronomy, Purdue University

Rachel Stevens: Purdue Agronomy Center for Research and Education (ACRE)

Aaron Kult: Purdue Agronomy Center for Research and Education (ACRE)

Evan Bossung: Purdue Agronomy Center for Research and Education (ACRE)

Study Location: Purdue Agronomy Center for Research and Education, West Lafayette, IN.

Soil Type: Chalmers silty clay loam (0–2% slope), Toronto-Millbrook complex (0–2% slope)

Planting Date: May 04, 2024 | **Harvest Date:** October 07, 2024

Corn Hybrid: Pioneer P1108Q | **Corn Seeding Rate:** 32,000 and 38,000 seeds/ac

Corn Nitrogen (N) Fertilizer Rate and Source: Nitrogen fertilizer was applied at a total rate of 180 lbs N/ac and included 40 lbs N/ac at planting (2x2 starter) across all treatment. See treatments below for in-season sidedress and fertigation fertilizer application specifics.

Previous Crop: Soybean | **Tillage:** Conventional

Study Replications: 4

RESEARCH TRIAL OVERVIEW:

This research trial was established at the Agronomy Center for Research and Education (ACRE) in Tippecanoe County, IN. This trial examined corn grain yield differences under conventional and intensive management with and without the use of recycled drainage water for application of irrigation and fertigation using subsurface-drip lines. The experimental design of this trial was a split-plot design with four replications. The main plot was water input and subplot was management. All plots measured 60 feet wide (24, 30-inch corn rows) x 65 feet long and the center four rows of each plot were harvested with a Wintersteiger Plot Combine equipped with a HarvestMaster weigh system and adjusted to 15.5% moisture for yield analysis.

RESEARCH TREATMENTS:

1. Control treatment (C) based on Purdue University seed rate (32K seeds/ac) and nitrogen (N) fertilizer rate recommendations (Camberato, J., R. L. Nielsen, and D. Quinn, 2022, *Nitrogen Management Guidelines for Corn in Indiana*, Purdue Univ. Extension, https://www.agry.purdue .edu/ext/corn/news/timeless/nitrogenmgmt.pdf; Nielsen, R. L., D. Quinn, and J. Camberato, 2022, *Optimum Plant Populations for Corn in Indiana*, Corny News Network. Purdue Univ. Extension, https://www.agry.purdue.edu/ext/corn/news/timeless/PlantPopulations.html).

2. C + Irrigation. Irrigation water was applied through a subsurface-drip application. Irrigation applications were decided daily based on soil moisture levels.

3. C + Fertigation consisted of 20 lbs N/ac applied as UAN (28-0-0) and an additional 2 lbs S/ac applied as ATS (12-0-0-26S) injected through the subsurface drip lines with irrigation. The applications were made at the V12, R1, and R2 growth stages.

4. Intensive Management (IM). Seeding rate of 38,000 seeds/ac, multiple in-season N fertilizer application [starter N (2x2) + V5 sidedress N (60% remaining N rate) + V10-12 growth stage sidedress N surface-banded with drop tubes on a sprayer (40% remaining N rate), total N rate remained the same as other treatments], sulfur fertilizer (5.2 gallons/ac as ammonium thiosulfate (ATS) at V5 sidedress), and foliar fungicide at the R1 growth stage (mefentrifluconazole, pyraclostrobin, Veltyma, 10 oz/ac).

5. Intensive Management (IM) + Irrigation. Treatment contained all intensive applications above combined with irrigation presence.

6. Intensive Management (IM) + Fertigation. Fertigation consisted of 20 lbs N/ac applied as UAN (28-0-0) and an additional 2 lbs S/ac applied as ATS (12-0-0-26S) injected through the subsurface drip lines with irrigation. The applications were made at the V12, R1, and R2 growth stages. Additionally, foliar fungicide at the R1 growth stage was applied (mefentrifluconazole, pyraclostrobin, Veltyma, 10 oz/ac).

RESULTS:

TABLE 15. *Corn grain moisture and grain yield differences observed from applied treatments in 2024. West Lafayette, IN.*

TREATMENT	GRAIN MOISTURE	GRAIN YIELD
	--- % ---	-- bu/ac --
Control (C)	17.9 c*	270.0 d
C + Irrigation	18.0 c	274.4 cd
C + Fertigation	17.9 c	268.6 d
Intensive (IM)	18.8 b	279.2 bc
IM + Irrigation	19.1 ab	283.0 b
IM + Fertigation	19.4 a	296.8 a
Pr>F	0.001	0.001

* Mean values that do not contain the same corresponding letter and are within the same column are determined statistically different ($P < 0.1$).

SUMMARY (TAKE-HOME POINTS):

- Intensive management treatments increased grain moisture at the end of the season (Table 15).
- Intensive management treatments plus the use of fertigation produced the highest yield in comparison to the control in 2024. In addition, the combination of intensive management plus fertigation resulted in a higher yield, whereas the control plus fertigation did not improve yield compared to the control.
- This research trial experienced high levels of tar spot in 2024. The yield increase due to the inclusion of intensive management was likely driven by the fungicide application.

- Preliminary results suggest irrigation/fertigation from recycled drainage water applied through subsurface-drip lines have the potential to increase corn yield by supplementing crop water needs during the season if there are dry conditions. Meanwhile, intensive management through higher seeding rates and the additional application of sulfur and fungicide could help to keep consistently higher yields when compared to the standard management treatment.

CORN RESPONSE TO FLUTRIAFOL FUNGICIDE APPLICATION TIMING AND PLACEMENT (ACRE)

Daniel Quinn: Department of Agronomy, Purdue University
Rachel Stevens: Purdue Agronomy Center for Research and Education (ACRE)
Aaron Kult: Purdue Agronomy Center for Research and Education (ACRE)
Evan Bossung: Purdue Agronomy Center for Research and Education (ACRE)

Study Location: Purdue Agronomy Center for Research and Education, West Lafayette, IN
Soil Type: Drummer silty clay loam (0–2% slope)
Planting Date: May 03, 2024 | **Harvest Date:** Oct 07, 2024
Corn Hybrid: Pioneer P1108Q | **Corn Seeding Rate:** 32,000 seeds/ac
Corn Nitrogen (N) Fertilizer Rate and Source: 40 lbs N/ac as 28% UAN in a 2x2 starter + 140 lbs N/ac as 28% UAN either coulter-injected as a sidedress application between the corn rows or surface-banded next to the corn rows at growth stage V5. Sidedress N application method was dependent on required Xyway LFR application method.
Previous Crop: Soybean | **Tillage:** Spring Vertical Tillage
Study Replications: 5

RESEARCH TRIAL OVERVIEW:

This research trial was established at the Agronomy Center for Research and Education (ACRE) in Tippecanoe County, IN. The trial examined corn yield response to Xyway LFR fungicide applied via three different application methods. In all application methods, Xyway LFR was tank-mixed and applied with 28% UAN. The trial was designed as a randomized complete block design with four treatments and five replications. Plots measure 15 feet wide (six 30-inch rows) by 40 feet long and the center two rows were harvested using a Wintersteiger Plot Combine equipped with a HarvestMaster weigh system and adjusted to 15.5% moisture for yield analysis.

RESEARCH TREATMENTS:

1. Nontreated control (no fungicide applied)
2. Xyway LFR (15 oz/ac) applied as a 2x2 starter application at planting
3. Xyway LFR (15 oz/ac) surface-banded next to the row (NR; next to row) at the V6 growth stage
4. Xyway LFR (15 oz/ac) coulter-injected band between the corn rows (BR; between rows) at the V6 growth stage

RESULTS:

TABLE 16. *Corn grain moisture (%) and yield (bu/ac) in response to Xyway LFR placement. West Lafayette, IN 2024.*

XYWAY LFR PLACEMENT	GRAIN MOISTURE	GRAIN YIELD
	---- % ----	---- bu/ac ----
Nontreated	16.6 a*	250.1 a
2x2 (At-planting)	16.4 a	254.1 a
NR (Next to row @ V5)	16.6 a	252.8 a
BR (Between rows @ V5)	16.5 a	253.7 a
Pr > F	0.677	0.917

* Mean values that do not contain the same letter are determined statistically different from each other ($P < 0.1$).

TABLE 17. *Corn ear leaf disease severity (R4 and R5 growth stage) in response to Xyway LFR placement. West Lafayette, IN 2024.*

XYWAY LFR PLACEMENT	R4 EAR LF	R5 EAR LF	R5 EAR LF
	--- % tar spot ---	--- % tar spot ---	-- % gray leaf spot --
Nontreated	19.3 a*	23.3 a	0.7 a
2x2 (At-planting)	11.2 b	14.8 b	0.6 ab
NR (Next to row @ V5)	12.2 b	14.7 b	0 b
BR (Between rows @ V5)	11.2 b	15.3 b	0 b
Pr > F	0.050	0.019	0.001

* Mean values that do not contain the same letter are determined statistically different from each other ($P < 0.1$).

SUMMARY (TAKE-HOME POINTS):

- All flutriafol treatments examined reduced growth stage R4 and R5 ear leaf tar spot disease severity and R5 ear leaf gray leaf spot severity (Table 17).
- Despite foliar disease severity reductions, no grain yield differences were observed across treatments (Table 16), which may indicate that foliar disease severity reductions were not significant enough to translate to grain yield improvement.

CORN YIELD AND GRAIN FILL RESPONSE TO FOLIAR FUNGICIDE AND FOLIAR SMARTKB APPLICATIONS AT THE R1 GROWTH STAGE (ACRE)

Daniel Quinn: Department of Agronomy, Purdue University
Rachel Stevens: Purdue Agronomy Center for Research and Education (ACRE)
Aaron Kult: Purdue Agronomy Center for Research and Education (ACRE)
Evan Bossung: Purdue Agronomy Center for Research and Education (ACRE)

Study Location: Purdue Agronomy Center for Research and Education, West Lafayette, IN
Soil Type: Raub-Brenton silt loam–silty clay loam (0–1% slope), Drummer silty clay loam (0–2% slope)
Planting Date: May 03, 2024 | **Harvest Date:** Oct 05, 2024
Corn Hybrid: Pioneer P1108Q | **Corn Seeding Rate:** 32,000 seeds/ac
Corn Nitrogen (N) Fertilizer Rate and Source: 40 lbs N/ac as 28% UAN in a 2x2 starter + 140 lbs N/ac as 28% UAN coulter-injected as a sidedress application between the corn rows.
Previous Crop: Soybean | **Tillage:** Spring Vertical Tillage
Study Replications: 5

RESEARCH TRIAL OVERVIEW:

A field research trial was conducted at the Purdue Agronomy Center for Research and Education (ACRE) in Tippecanoe County, IN. The research trial examined corn yield response to Smart KB (16% soluble potash + 2.5% boron) and Delaro Complete fungicide foliar applied at the R1 growth stage. The trial was designed as a randomized complete block design with four treatments and five replications. Plots measure 15 feet wide (six 30-inch rows) by 40 feet long and the center two rows were harvested using a Wintersteiger Plot Combine equipped with a HarvestMaster weigh system and adjusted to 15.5% moisture for yield analysis.

RESEARCH TREATMENTS:

1. Nontreated Control
2. Delaro Complete (10 oz/ac at R1, application rate of 20 gal/ac)
3. Smart KB (2 qt/ac at R1, application rate of 20 gal/ac)
4. Delaro Complete (10 oz/ac) + Smart KB (2 qt/ac) both tank-mixed and applied at R1 with an application rate of 20 gal/ac

RESULTS:

TABLE 18. *Corn grain moisture (%) and yield (bu/ac) in response to R1 growth stage foliar-applied fungicide and fertilizer products. West Lafayette, IN 2024.*

FOLIAR TREATMENT	GRAIN MOISTURE	GRAIN YIELD
	---- % ----	---- bu/ac ----
Nontreated Control	18.3 b*	257.3 b
Delaro Complete (10 oz/ac at R1)	19.1 a	265.1 ab
Smart KB (2 qt/ac at R1)	18.9 ab	264.4 ab
Delaro Complete + SmartKB	18.9 ab	271.7 a
Pr > F	*0.024*	*0.031*

* Mean values that do not contain the same letter and are within the same column are determined statistically different from each other ($P < 0.1$).

Kernel Sampling Methodology: A total of two predetermined ears were sampled from rows 2 and 5 of each individual plot (four total ears) each week, beginning at one week after silking (July 22, 2024). A total of 10 kernels per sampled ear were extracted from the middle of the ear and dried to a constant moisture to obtain dry weight per kernel.

TABLE 19. *Corn grain fill duration and maximum kernel weight (mg/kernel) in response to R1 foliar-applied fungicide and fertilizer products. West Lafayette, IN 2024.*

FOLIAR TREATMENT	DAYS AFTER SILKING	MAX DRY KERNEL WEIGHT	R²	P-VALUE
	---- days ----	--- mg/kernel ---		
Nontreated	59.1*	453*	0.93	*0.001*
Delaro Complete (10 oz/ac @ R1)	60.4	454	0.95	*0.001*
Smart KB (2 qt/ac @ R1)	59.0	460	0.91	*0.001*
Delaro Complete + SmartKB	63.2	473	0.94	*0.001*

* Mean values were acquired using quadratic plateau regression analysis performed using the "easynls" package in R.

SUMMARY (TAKE-HOME POINTS):

- The inclusion of the R1 growth stage foliar fungicide + foliar fertilizer containing potassium and boron produced the largest grain yield difference in comparison to the nontreated control (Table 18).
- In addition, the foliar fungicide + foliar fertilizer treatment resulted in the highest grain fill duration and maximum kernel dry weight achieved in comparison to all other treatments examined, which likely contributed to the grain yield differences observed (Table 19).

CORN RESPONSE TO ASYMBIOTIC N-FIXING BIOINOCULANT PRODUCTS (ACRE)

Narciso Zapata: Department of Agronomy, Purdue University

Roland Wilhelm: Department of Agronomy, Purdue University

Daniel Quinn: Department of Agronomy, Purdue University

Rachel Stevens: Purdue Agronomy Center for Research and Education (ACRE)

Aaron Kult: Purdue Agronomy Center for Research and Education (ACRE)

Evan Bossung: Purdue Agronomy Center for Research and Education (ACRE)

Study Location: Purdue Agronomy Center for Research and Education, West Lafayette, IN.

Soil Type: Drummer silty clay loam (0–2% slope), Raub-Brenton complex silt loam–silty clay loam (0–1% slope)

Planting Date: May 08, 2024 | **Harvest Date:** October 10, 2024

Corn Hybrid: Pioneer P1108Q | **Corn Seeding Rate:** 32,000 seeds/ac

Corn Nitrogen (N) Fertilizer Rate and Source: 0, 60, 120, and 180 lbs N/ac, all nitrogen fertilizer rate applications were applied as 28% UAN coulter-injected as a sidedress application between the corn rows at growth stage V4. No starter fertilizer used in this trial.

Previous Crop: Soybean | **Tillage:** Conventional

Study Replications: 5

RESEARCH TRIAL OVERVIEW:

A field research trial was established at the Purdue Agronomy Center for Research and Education (ACRE) in Tippecanoe County, IN. The research trial examined corn yield response to different biological products application and N rates to analyze the impact of the interaction between both factors. The trial was designed as a randomized complete block design with 20 treatments and five replications. Plots measured 15 feet wide (six 30-inch corn rows) by 40 feet long and the center four rows were harvested using a Wintersteiger Plot Combine equipped with a HarvestMaster weigh system and adjusted to 15.5% moisture for yield analysis.

RESEARCH TREATMENTS:

BIOLOGICAL PRODUCT:

1. No biological
2. Envita SC foliar applied at V6 (Azotic NA)
3. UtrishaN foliar applied at V6 (Corteva Agrisciences)
4. Proven40 on-seed treatment at planting (PivotBio)
5. Source foliar applied at V6 (Sound Ag)

NITROGEN FERTILIZER RATE (COULTER-INJECTED BETWEEN THE CORN ROWS AT V5, NO STARTER APPLIED):

1. 0 lbs N/ac
2. 60 lbs N/ac
3. 120 lbs N/ac
4. 180 lbs N/ac

RESULTS:

TABLE 20. *Mean grain yield (bu/ac) differences observed across biological products and nitrogen (N) fertilizer application rate. West Lafayette, IN 2024.*

NITROGEN RATE	BIOLOGICAL PRODUCT	GRAIN YIELD
-- lbs N/ac --		-- bu/ac --
0	None	192.1 g*
	Envita SC	196.2 g
	Proven40 OS	190.9 g
	Source	190.8 g
	Utrisha N	191.9 g
60	None	253.1 ef
	Envita SC	253.6 ef
	Proven40 OS	265.4 d
	Source	262.6 de
	Utrisha N	248.9 f
120	None	285.7 c
	Envita SC	282.9 c
	Proven40 OS	290.7 bc
	Source	288.6 c
	Utrisha N	286.8 c
180	None	298.6 ab
	Envita SC	307.5 a
	Proven40 OS	299.6 ab
	Source	303.6 a
	Utrisha N	299.5 ab

* Mean values that do not contain the same corresponding letter are determined statistically different ($P < 0.1$).

TABLE 21. *Corn grain yield (bu/ac) response to biological products. Data combined across all applied nitrogen fertilizer rates in the research trial. West Lafayette, IN 2024.*

TREATMENT DESCRIPTION	GRAIN YIELD
	---- bu/ac ----
Nontreated	257.4 ab*
Source—Foliar Applied at V6	261.4 ab
Envita SC—Foliar Applied at V6	260.1 ab
Utrisha-N—Foliar Applied at V6	256.8 b
Proven 40 OS—Seed Trt	261.6 a

* Mean values that contain dissimilar letters and are in the same column are determined significantly different from each other ($P < 0.1$).

SUMMARY (TAKE-HOME POINTS):

- Corn yield was significantly ($P < 0.1$) increased with increases in N fertilizer rates. However, preliminary results suggest examined biological products did not improve corn yield at any of the examined N rates and no interaction between biological presence and corn yield response to N fertilizer rate was observed (Tables 20 and 21).

- Overall, results suggest application of asymbiotic N-fixing bioinoculants do not improve corn yield or reduce synthetic N fertilizer requirements for corn at this location. However, analysis of additional locations is still ongoing and trials will be replicated on more locations in 2025.

CORN RESPONSE TO NITROGEN-FIXING BIOLOGICALS AND SULFUR IN A CEREAL RYE COVER CROP SYSTEM (ACRE)

Daniel Quinn: Department of Agronomy, Purdue University

Narciso Zapata: Department of Agronomy, Purdue University

Rachel Stevens: Purdue Agronomy Center for Research and Education (ACRE)

Aaron Kult: Purdue Agronomy Center for Research and Education (ACRE)

Evan Bossung: Purdue Agronomy Center for Research and Education (ACRE)

Study Location: Purdue Agronomy Center for Research and Education, West Lafayette, IN

Soil Type: Raub-Brenton complex (0–1% slope), Drummer silty clay loam (0–2% slope)

Planting Date: May 03, 2024 | **Harvest Date:** Oct. 09, 2024

Corn Hybrid: Pioneer P1108Q | **Corn Seeding Rate:** 32,000 seeds/ac

Corn Nitrogen (N) Fertilizer Rate and Source: See treatments below

Previous Crop: Soybean | **Tillage:** No-till

Study Replications: 8

RESEARCH TRIAL OVERVIEW:

A field research trial was established at the Purdue Agronomy Center for Research and Education (ACRE) in Tippecanoe County, IN. The research trial examined corn response to Proven40OS and V5 coulter-injected sidedress sulfur in a cereal rye cover crop system. The cereal rye was fall drill-seeded at 45 lbs/ac and chemically terminated three weeks prior to corn planting. The trial was designed as a split-plot design with eight treatments and eight replications. Plots measured 15 feet wide (six 30-inch rows) by 40 feet long and the center two rows were harvested using a Wintersteiger Plot Combine equipped with a HarvestMaster weigh system and adjusted to 15.5% moisture for yield analysis.

RESEARCH TREATMENTS:

1. No cover crop, 2x2 starter with 40 lbs N/ac as 28% UAN followed by V5 sidedress with 160 lbs N/ac as UAN

2. No cover crop, 2x2 starter with 40 lbs N/ac as UAN followed by sidedress with 153 lbs N/ac UAN plus 7 lbs N/ac and 15 lbs S/ac as Ammonium Thiosulfate (ATS) 12-0-0-26S

3. No cover crop, Proven40OS seed treatment, followed by sidedress 160 lbs N/ac UAN. No starter N included with Proven40OS seed treatment.

4. No cover crop, Proven40OS, followed by sidedress 143 lbs N/ac as UAN plus 7 lbs N/ac and 15 lbs S/ac as ATS

5. Cereal rye cover crop, 2x2 stater 40 lbs N/ac as UAN followed by sidedress with 160 lbs N/ac as UAN

6. Cereal rye cover crop, 2x2 stater 40 lbs N/ac as UAN followed by sidedress with 143 lbs N/ac as UAN plus 7 lbs N/ac as ATS

7. Cereaı rye cover crop, Proven40OS seed treatment, followed by sidedress 160 lbs N/ac UAN. No starter N included with Proven40OS seed treatment.

8. Cereal rye cover crop, Proven40OS, followed by sidedress 143 lbs N/ac as UAN plus 7 lbs N/ac and 15 lbs S/ac as ATS

RESULTS:

TABLE 22. *Corn yield response to seed-applied Proven40OS, Ammonium Thiosulfate (ATS) applied with sidedress UAN at the V5 growth stage, and different nitrogen fertilizer rates when following a cereal rye cover crop. Cereal rye was fall drill-seeded at 40 lbs/ac and terminated two weeks prior to corn planting. West Lafayette, IN 2024. No biological treatments had 40 lbs N/ac applied in a 2x2 starter at planting where Proven40 OS treatments did not.*

RYE COVER CROP	BIOLOGICAL PRODUCT	NITROGEN RATE	GRAIN YIELD
		---- lbs N/ac ----	---- bu/ac ----
No	None	200	304.2 a*
	Proven40OS	160	295.4 b
	ATS (V5)	200	305.6 a
	Proven40OS + ATS	160	294.1 b
Yes	None	200	271.2 d
	Proven40OS	160	270.1 d
	ATS (V5)	200	284.4 c
	Proven40OS + ATS	160	268.8 d

* Mean values that do not contain the same corresponding letter are determined statistically different ($P < 0.1$).

SUMMARY (TAKE-HOME POINTS):

- Without the presence of a cereal rye cover crop, the inclusion of Proven40OS or ATS did not improve yield beyond the nontreated control (Table 22). Grain yield was also reduced with the inclusion of Proven40OS, which was likely due to the reduced N fertilizer rate applied.

- With the inclusion of a cereal rye cover crop, corn grain yield was improved with the application of ATS combined with UAN at V5 sidedress (Table 22). In addition, the inclusion of Proven40OS with a lower N fertilizer rate applied was statistically nonsignificant from the nontreated control with a higher N fertilizer rate.

- Nitrogen and S deficiencies were observed in corn following the cereal rye cover crop, likely due to high carbon levels and N and S immobilization. These deficiencies likely drove the observed responses to the ATS and Proven40OS in the cover crop system.

- Research trial will be expanded in 2025 for additional data collection and understanding of results.

CORN RESPONSE TO LONG-TERM TILLAGE IN CONTINUOUS CORN (C-C) AND CORN AFTER SOYBEAN (C-SB) ROTATIONS (ACRE)

Raziel Ordóñez: Department of Agronomy, Purdue University

Tony Vyn: Department of Agronomy, Purdue University

Daniel Quinn: Department of Agronomy, Purdue University

Rachel Stevens: Purdue Agronomy Center for Research and Education (ACRE)

Aaron Kult: Purdue Agronomy Center for Research and Education (ACRE)

Evan Bossung: Purdue Agronomy Center for Research and Education (ACRE)

Study Location: Purdue Agronomy Center for Research and Education, West Lafayette, IN

Soil Type: Chalmers silty clay loam (0–2% slope)

Planting Date: May 20, 2024 | **Harvest Date:** Oct 11, 2024

Corn Hybrid: Pioneer 14830Q | **Corn Seeding Rate:** 35,000 seeds/ac

Corn Nitrogen (N) Fertilizer Rate and Source: 40 lbs N/ac as 28% UAN in a 2x2 starter + 160 lbs N/ac as 28% UAN coulter-injected as a sidedress application between the corn rows.

Previous Crop: Soybean, Corn | **Tillage:** Fall Moldboard Plow, Chisel Plow and Interrow Strip-Till C-C and On-row Strip-till (C-B); Fall Moldboard, Chisel and Strip Till occurred 11/12/2023; Spring Seed Bed Preparation Tillage occurred 5/13/2024 with a 1x field cultivator pass

Study Replications: 4

RESEARCH TRIAL OVERVIEW:

This research is a continuation of the Purdue Long-Term Tillage Study started in 1975. Corn yield response was evaluated in both continuous corn (C-C) and corn after soybeans (C-S) following fall moldboard plow, chisel plow, strip-tillage, and no-till tillage systems. The moldboard and chisel plow treatments also received one medium to shallow spring tillage pass of a field cultivator for spring seedbed preparation. The study was a split-plot design with eight treatments and four replications. Plots are 30 feet (12, 30-inch rows) wide by 150 feet long and the center four rows were harvested using a Wintersteiger Plot Combine equipped with a HarvestMaster weigh system and adjusted to 15.5% moisture for yield analysis.

RESEARCH TREATMENTS:

1. Continuous Corn—Mold Board Plow
2. Continuous Corn—Chisel Plow
3. Continuous Corn—Strip-Till
4. Continuous Corn—No-Till
5. Corn after Soybeans—Moldboard Plow
6. Corn after Soybeans—Chisel Plow
7. Corn after Soybeans—Strip-Till
8. Corn after Soybeans—No-Till

RESULTS:

TABLE 23. *Plant density and grain yield under rotation and within tillage systems. West Lafayette, IN 2015–2024.*

TILLAGE SYSTEM	PLANT DENSITY (PLANTS/AC)		GRAIN YIELD (BU/AC)	
	CONTINUOUS CORN	CORN-SOYBEAN ROTATION	CONTINUOUS CORN	CORN-SOYBEAN ROTATION
Plow	33700 abc*	34100 a	238.7 b	251.3 a
Chisel	33200 bc	33800 ab	224.1 c	246.3 ab
Strip	33400 abc	33400 abc	229.4 c	252.8 a
No-Till	33000 c	33800 ab	229.8 c	248.4 a
Average	**33300 B**	**33800 A**	**229.5 B**	**249.7 A**

* Mean values that do not contain the same letter and are within the same column are determined statistically different from each other ($P < 0.1$)

SUMMARY (TAKE-HOME POINTS):

- Results from the past 10 years of crop rotation and tillage intensities research showed significant differences between corn following soybean versus corn following corn (249.7 bu/ac versus 229.5 bu/ac, respectively).
- Yields within rotation ranged from 246.4 to 252.8 bu/ac, and in corn following corn yield values ranged from 224.1 to 238.7 bu/ac.
- While yields were nearly identical across tillage systems under rotation, values under continuous corn were more variable, with Moldboard Plow showing a greater value compared to the other tillage intensities.
- The overall yield advantage due to rotation over the last decade was 8.8% (Table 23). This yield advantage due to rotation was consistent across tillage intensities, ranging from 5.3% to 10.2%. The greater yield advantage occurred under Chisel, while the smallest was observed under Moldboard Plow. These findings indicate that crop rotation plays a significant role in yield improvements, whereas under continuous corn, tillage intensity is important to mitigate yield losses (Table 23).
- Stem count was statistically higher by 1.5% under rotation compared to corn on corn (Table 23).
- Crop rotation reduced corn plant mortality by up to 2.4%, particularly under No-Till systems in corn after corn production system. Significant differences were observed for No-Till under continuous corn, while other systems showed no statistical differences.
- A key finding is that higher yields under Moldboard Plow may be attributed to improved seed placement, which ensures the targeted plant population. In contrast, the lower yields observed under No-Till after corn might be linked to higher early season plant mortality rates (Table 1).

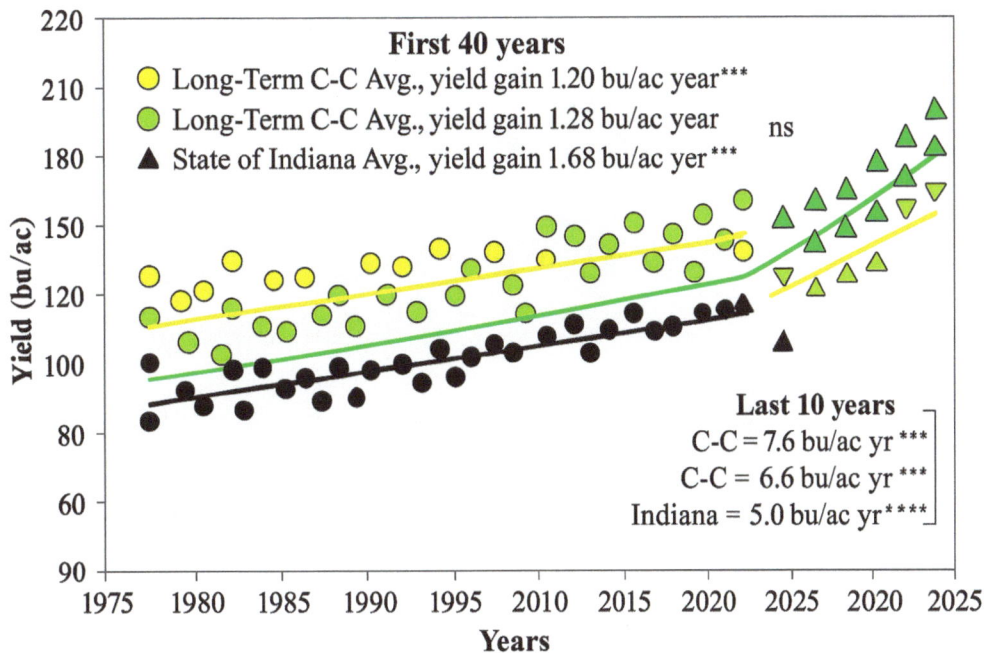

FIGURE 4. Corn yield gains over time for continuous corn, corn after soybean, and the State of Indiana.

SUMMARY (TAKE-HOME POINTS):

- Corn yield gains have steadily increased over the past 50 years in both corn following soybean and continuous corn systems (Figure 4). Over the first 40 years of experimentation, corn following soybean showed a slightly higher yield gain than corn on corn, although both were below the state average (Figure 4).

- In the past 10 years, the rate of yield increase has accelerated, with corn on corn yielding 7.6 bu/ac yr⁻¹ and corn following soybean yielding 6.6 bu/ac yr⁻¹. Importantly, these values exceeded the state average of 5.0 bu/ac yr⁻¹. The increased yield gains values over the past 10 years are likely attributed to the efforts of breeding programs in consistently developing hybrids that are more tolerant to abiotic and biotic stresses compared to those in the past (Figure 4).

POPCORN AND DENT CORN RESPONSE TO NITROGEN RATE (ACRE)

Narciso Zapata: Department of Agronomy, Purdue University

Daniel Quinn: Department of Agronomy, Purdue University

Rachel Stevens: Purdue Agronomy Center for Research and Education (ACRE)

Aaron Kult: Purdue Agronomy Center for Research and Education (ACRE)

Evan Bossung: Purdue Agronomy Center for Research and Education (ACRE)

Study Location: Purdue Agronomy Center for Research and Education, West Lafayette, IN.

Soil Type: Chalmers silty clay loam (0–2% slope)

Planting Date: May 08, 2024 | **Harvest Date:** October 10, 2024

Corn Hybrid: Weaver 2021 and Pioneer P1108Q | **Corn Seeding Rate:** 34,000 seeds/ac

Corn Nitrogen (N) Fertilizer Rate and Source: 0, 60, 120, 180, and 240 lbs N/ac. All nitrogen fertilizer rate applications were applied as 28% UAN coulter-injected as a sidedress application between the corn rows at growth stage V4. No starter fertilizer was applied in this trial.

Previous Crop: Soybean | **Tillage:** Conventional

Study Replications: 5

RESEARCH TRIAL OVERVIEW:

A field research trial was established at the Purdue Agronomy Center for Research and Education (ACRE) in Tippecanoe County, IN. The research trial assessed corn yield differences when utilizing different N fertilizer application rates in popcorn and dent corn. The trial was designed as a randomized complete block design with 10 treatments and five replications. Plots are 30 feet (12, 30-inch rows) wide by 40 feet long and the center four rows were harvested using a Wintersteiger Plot Combine equipped with a HarvestMaster weigh system and adjusted to 15.5% moisture for yield analysis.

RESEARCH TREATMENTS:

CORN TYPE:

1. Dent Corn
2. Popcorn

NITROGEN FERTILIZER RATE (COULTER-INJECTED BETWEEN THE CORN ROWS AT V5, NO STARTER APPLIED):

1. 0 lb N/ac
2. 60 lb N/ac
3. 120 lb N/ac
4. 180 lb N/ac
5. 240 lb N/ac

RESULTS:

TABLE 24. *Quadratic plateau regression analysis parameters and calculated agronomic optimum nitrogen fertilizer rates (AONR) and yields obtained at the calculated AONR values for popcorn and dent corn. West Lafayette, IN 2024.*

CORN TYPE	MODEL†	REGRESSION PARAMETERS			AONR	YAONR†
		INTERCEPT	LINEAR COEFFICIENT	QUADRATIC COEFFICIENT		
					lbs N ac⁻¹	lbs ac⁻¹
Dent Corn	QP	188.9	1.00	-0.002	**195.2**	287.1
Popcorn	QP	5035.9	17.17	-0.058	**146.1**	6289.6

† QP, quadratic-plateau regression model; YAONR, popcorn grain yield at the agronomic optimum N rate (AONR).

SUMMARY (TAKE-HOME POINTS):

- Agronomic optimum N rate (AONR) for popcorn was 146.1 lbs N/ac to produce a grain yield of 6290 lbs/ac (Table 24). In comparison, the AONR for dent corn grown in the same environment, same field, and using the same management practices was 195.2 lbs N/ac to produce a maximum grain yield of 287 bu/ac.
- Results suggest at this location in 2024, popcorn required ~50 lbs N/ac less to produce maximum grain yield in comparison to dent corn.

EVALUATION OF CORN HYBRID RESPONSE TO LEAF DEFOLIATION SEVERITY AND TIMING (ACRE)

Daniel Quinn: Department of Agronomy, Purdue University
Rachel Stevens: Purdue Agronomy Center for Research and Education (ACRE)
Aaron Kult: Purdue Agronomy Center for Research and Education (ACRE)
Evan Bossung: Purdue Agronomy Center for Research and Education (ACRE)

Study Location: Purdue Agronomy Center for Research and Education, West Lafayette, IN
Soil Type: Drummer silty clay loam (0–2% slope)
Planting Date: May 09, 2024 | **Harvest Date:** Oct. 09, 2024
Corn Hybrid: Dekalb DKC62-70; Preceon PR111-20SSC, PR108-20SSC
Corn Seeding Rate: 34,000 seeds/ac (DKC62-70) and 38,000 seeds/ac (PR111-20SSC and PR108-20SSC).
Corn Nitrogen (N) Fertilizer Rate and Source: 40 lbs N/ac as 28% UAN in a 2x2 starter + 140 lbs N/ac as 28% UAN coulter-injected as a sidedress application between the corn rows.
Previous Crop: Soybean | **Tillage:** Spring Vertical Tillage
Study Replications: 4

RESEARCH TRIAL OVERVIEW:

A field research trial was established at the Purdue Agronomy Center for Research and Education (ACRE) in Tippecanoe County, IN. The trial evaluated corn hybrid response of one full-stature corn hybrid and two short-stature corn hybrids at four defoliation amounts at three different growth stages. A blanket foliar fungicide application of Delaro Complete was applied at the R2 growth stage. The trial was a randomized completed block design with four replications. Plot size measured 15 feet wide (six 30-inch corn rows) x 40 feet long. Grain yield was harvested from the center two rows using a Wintersteiger Plot Combine equipped with a HarvestMaster weigh system and adjusted to 15.5% moisture.

RESEARCH TREATMENTS:

CORN HYBRID TYPE:

1. Full-stature hybrid DKC62-70
2. Short-stature hybrid PR111-20SSC
3. Short-stature hybrid PR108-20SS

DEFOLIATION AMOUNT:

1. 0% defoliation
2. 33% defoliation
3. 66% defoliation
4. 99% defoliation

DEFOLIATION GROWTH STAGE TIMING:

1. V8
2. V13
3. R1

RESULTS:

TABLE 25. *Analysis of variance (ANOVA) for the dependent variable of grain yield (bu/ac) and the main effects of hybrid, defoliation timing, and defoliation percentage, and their subsequent interactions.*

EFFECT	NM DF	DEN DF	F VALUE	PR > F
Hybrid	2	207	17.34	<0.001
Defoliation Timing	2	207	69.18	<0.001
Defoliation %	2	207	240.66	<0.001
Hybrid*Defoliation Timing	4	207	1.43	0.225
Hybrid*Defoliation %	4	207	0.47	0.755
Defoliation Timing * Defoliation%	4	207	40.08	<0.001
Hybrid*DefoliationStage*Defoliation%	8	207	0.32	0.959

TABLE 26. *Corn leaf area index (LAI), grain moisture, grain yield, and lodging incidence in response to hybrid, defoliation timing, and defoliation percentage. West Lafayette, IN.*

HYBRID	DEFOLIATION GROWTH STAGE	DEFOLIATION %	GRAIN MOISTURE	GRAIN YIELD	LODGING
			-- % --	-- bu/ac --	-- % --
PR108-20SSC	UTC	0	17.5 a	271.9 ab*	0
	V8	33	17.1 a	272.8 ab	0
		66	17.4 a	264.9 ab	0
		99	17.3 a	251.4 bc	0
	V13	33	17.5 a	273.4 a	0
		66	17.3 a	254.1 abc	0
		99	15.9 b	153.7 d	0
	R1	33	17.3 a	272.8 ab	0
		66	17.3 a	242.9 c	0
		99	14.6 c	127.1 e	0
PR111-20SSC	UTC	0	18.1 a	271.6 a	0
	V8	33	18.4 a	267.5 ab	0
		66	18.3 a	247.5 bcd	0
		99	18.4 a	242.6 cd	0
	V13	33	18.3 a	258.4 abc	0
		66	18.1 a	243.3 cd	0
		99	16.7 b	138.9 e	0
	R1	33	18.1 a	267.1 ab	0
		66	18.0 a	234.1 d	0
		99	15.1 c	135.8 e	0
DKC62-70	UTC	0	17.1 ab	293.3 ab	0
	V8	33	16.7 bcd	304.1 a	0
		66	17.1 abc	291.1 ab	0
		99	17.5 a	284.1 bc	0
	V13	33	17.1 abc	295.3 ab	0
		66	17.1 ab	270.8 bc	0
		99	16.1 d	157.5 d	21.2
	R1	33	17.1 abc	287.1 abc	0
		66	16.3 cd	263.3 c	0
		99	13.4 e	133.2 d	0

* Mean values that contain dissimilar letters and are in the same column and within the same hybrid are determined significantly different from each other ($P < 0.1$).

SUMMARY (TAKE-HOME POINTS):

- No significant (P < 0.1) grain yield interactions were observed between hybrid type and defoliation timing or defoliation percentage, which indicates similar yield responses to leaf defoliation across examined hybrids (Tables 25 and 26).
- Lodging was only observed in the full-stature corn hybrid at the V13, 99% leaf defoliation treatment (Table 26).
- Leaf defoliation during the V8 growth stage had the biggest impacts on overall plant height and ear height across hybrids.
- Results suggest potentially higher corn yield tolerances to severe levels of defoliation at late-vegetative (V13) and reproductive (R1) growth stages. Therefore, total yield loss estimates from different leaf defoliation levels and timings may need to be revisited with newer corn genetics.

PINNEY PURDUE AGRICULTURAL CENTER (PPAC)

CORN RESPONSE TO KORN-KALI AND KORN-KALI + B
AT VARIOUS K RATES AND FERTILIZER COMBINATIONS (PPAC)

Daniel Quinn: Department of Agronomy, Purdue University

Stephen Boyer: Pinney Purdue Agricultural Center

Alex Leman: Pinney Purdue Agricultural Center

Study Location: Pinney Purdue Agricultural Center, Wanatah, IN

Soil Type: Sebewa shaly sand substratum (0–2% slope)

Planting Date: May 28, 2024 | **Harvest Date:** Oct. 23, 2024

Corn Hybrid: Pioneer P1108Q | **Corn Seeding Rate:** 32,000 seeds/ac

Corn Nitrogen (N) Fertilizer Rate and Source: 40 lbs N/ac as 28% UAN in a 2x2 starter at planting + 160 lbs N/ac as 28% UAN coulter-injected as a sidedress application between the corn rows at growth stage V5.

Previous Crop: Soybean | **Tillage:** Conventional Tillage

Study Replications: 5

RESEARCH TRIAL OVERVIEW:

A field research trial was established at the Pinney Purdue Agricultural Center (PPAC) in LaPorte County, IN. The trial examined Korn KALI (0-0-40-6-13S) and Korn KALI +B (0-0-40-6-13S-0.25 B) with and without other soil K and S fertilizers. The dry fertilizer treatments were broadcast applied (one day before planting) and incorporated. All combinations of K fertilizer products were applied to an equivalent rate of 100 lbs K/ac. The trial was designed as a randomized complete block design with seven treatments and five replications. Plot size measured 15 feet wide (six 30-inch corn rows) x 40 feet long and the center two rows were harvested using a Wintersteiger Plot Combine equipped with a HarvestMaster weigh system and adjusted to 15.5% moisture.

RESEARCH TREATMENTS:

1. Nontreated Control
2. Potash (0-0-60) applied at a rate equivalent to 100 lbs K/ac
3. Korn KALI (0-0-40-6-13S) at 50 lbs K/ac + Potash (0-0-60) at 50 lbs K/ac
4. Korn KALI +B (0-0-40-6-13S-0.25B) at 50 lbs K/ac + Potash (0-0-60) at 50 lbs K/ac
5. Korn KALI (0-0-40-6-13S) at 100 lbs K/ac
6. Korn KALI +B (0-0-40-6-13S-0.25B) at 100 lbs K/ac
7. K-Mag Standard (0-0-22-11-22S) at 59 lbs K/ac + Potash (0-0-60) at 41 lbs K/ac

RESULTS:

TABLE 27. *Corn plant stand (plants/ac), grain moisture (%) and yield (bu/ac) in response to fertilizer treatment. Wanatah, IN 2024.*

FERTILIZER TREATMENT	PLANT STAND	GRAIN MOISTURE	GRAIN YIELD
	-- plants/ac --	---- % ----	---- bu/ac ----
1. Nontreated (0 lbs K/ac)	31581 a*	16.9 a	254.9 a
2. Potash (0-0-60, 100 lbs K/ac)	31537 a	17.1 a	262.6 a
3. Korn-KALI (0-0-40-13S, 50 lbs K/ac) + Potash (50 lbs K/ac)	31537 a	17.0 a	260.2 a
4. Korn-KALI+B (0-0-40-13S, 50 lbs K/ac) + Potash (50 lbs K/ac)	31537 a	16.9 a	256.8 a
5. Korn-KALI (100 lbs K/ac)	31581 a	17.0 a	259.9 a
6. Korn-KALI + B (100 lbs K/ac)	31537 a	17.2 a	258.4 a
7. KMag (0-0-22-22S, 59 lbs K/ac) + Potash (41 lbs K/ac)	31712 a	17.1 a	264.2 a
P > F	0.3069	0.9107	0.8227

* Mean values that do not contain the same letter and are within the same column are determined statistically different from each other ($P < 0.1$).

SUMMARY (TAKE-HOME POINTS):

- No statistical plant stand, moisture, or yield differences were observed across treatments in this research trial (Table 27).
- Soil test potassium levels were all above critical at this research trial location.

CORN RESPONSE TO AT-PLANT IN-FURROW AND SEED-APPLIED FERTILIZER (PPAC)

Daniel Quinn: Department of Agronomy, Purdue University
Stephen Boyer: Pinney Purdue Agricultural Center
Alex Leman: Pinney Purdue Agricultural Center

Study Location: Pinney Purdue Agricultural Center, Wanatah, IN
Soil Type: Sebewa shaly sand substratum (0–2% slope)
Planting Date: May 28, 2024 | **Harvest Date:** Oct. 23, 2024
Corn Hybrid: Pioneer P1108Q | **Corn Seeding Rate:** 32,000 seeds/ac
Corn Nitrogen (N) Fertilizer Rate and Source: 200 lbs N/ac applied as 28% UAN and coulter-injected as a side-dress application at growth stage V5. No additional starter (2x2 or in-furrow) N fertilizer was applied in this trial.
Previous Crop: Soybean | **Tillage:** Conventional
Study Replications: 5

RESEARCH TRIAL OVERVIEW:

A field research trial was established at the Pinney Purdue Agricultural Center (PPAC) in LaPorte County, IN. The trial examined corn grain moisture and yield differences of in-furrow and seed-applied nutritional products from Brandt. The experimental design of this trial was a randomized complete block design with four replications. All plots measured 15 feet wide (six 30-inch corn rows) x 40 feet long. The center two rows of each plot were harvested with a Wintersteiger small-plot combine equipped with a HarvestMaster weigh system and adjusted to 15.5% moisture for yield analysis.

RESEARCH TREATMENTS:

1. Nontreated control (no in-furrow products applied)
2. Ammonium Polyphosphate (APP, 10-34-0, in-furrow at 5 gal/ac)
3. EnzUP Phosphorus DS only (in-furrow at 5 lbs/ac, dry soluble product)
4. APP (in-furrow at 2.5 gal/ac) + EnzUP Phosphorus DS (in-furrow at 2.5 lbs/ac). Rates were reduced for both products to apply same total P as treatments 2 and 3.
5. APP (in-furrow at 5 gal/ac) + EnzUP Zinc (in-furrow at 2 qt/ac)
6. APP (in-furrow at 5 gal/ac) + EnzUP Seedflow Zinc (seed-applied at 2 oz/80,000 seed unit)
7. EnzUP Phosphorus DS (in-furrow at 2.5 lbs/ac)

RESULTS:

TABLE 28. *Corn grain moisture (%) and yield (bu/ac) in response to in-furrow and seed-applied products. Wanatah, IN 2024.*

AT-PLANT TREATMENT	GRAIN MOISTURE	GRAIN YIELD
	---- % ----	---- bu/ac ----
Nontreated	18.9 a*	216.3 a
Ammonium Polyphosphate (10-34-0, 5 gal/ac)	19.1 a	223.3 a
EnzUP P DS (5 lbs/ac)	19.4 a	226.9 a
APP (2.5 gal/ac) + EnzUP P DS (2.5 lb/ac)	18.9 a	219.9 a
APP (5 gal/ac) + EnzUP Zn (1 qt/ac)	18.9 a	220.8 a
APP (5 gal/ac) + EnzUP SeedFlow Zn (1.3 oz/unit)	19.3 a	220.0 a
EnzUP P DS (2.5 lb/ac)	18.9 a	220.4 a
Pr > F	0.573	0.881

* Mean values that do not contain the same letter and are within the same column are determined statistically different from each other ($P < 0.1$).

SUMMARY (TAKE-HOME POINTS):

- No statistical plant stand, moisture, or yield differences were observed across treatments in this research trial (Table 28).
- Soil test phosphorus levels were all above critical at this research trial location.

CORN RESPONSE TO ASYMBIOTIC N-FIXING BIOINOCULANT PRODUCTS (PPAC)

Narciso Zapata: Department of Agronomy, Purdue University

Roland Wilhelm: Department of Agronomy, Purdue University

Daniel Quinn: Department of Agronomy, Purdue University

Stephen Boyer: Pinney Purdue Agricultural Center

Alex Leman: Pinney Purdue Agricultural Center

Study Location: Pinney Purdue Agricultural Center, Wanatah, IN

Soil Type: Sebewa loam (0–2% slope)

Planting Date: May 23, 2024 | **Harvest Date:** October 22, 2024

Corn Hybrid: Pioneer P1108Q | **Corn Seeding Rate:** 32,000 seeds/ac

Corn Nitrogen (N) Fertilizer Rate and Source: 0, 60, 120, 180, and 240 lbs N/ac. All nitrogen fertilizer rate applications were applied as 28% UAN coulter-injected as a sidedress application between the corn rows at growth stage V4. No starter fertilizer was used in this trial.

Previous Crop: Soybean | **Tillage:** Conventional

Study Replications: 5

RESEARCH TRIAL OVERVIEW:

A field research trial was established at the Pinney Purdue Agricultural Center (PPAC) in Porter County, IN. The research trial examined corn yield response to different biological products application and N rates to analyze the impact of the interaction between both factors. The trial was designed as a randomized complete block design with 16 treatments and five replications. The research trial was performed in a field that maintained a corn-soybean rotation. Plots measured 15 feet wide (six 30-inch corn rows) by 40 feet long and the center four rows were harvested with a small-plot combine and adjusted to 15.5% moisture for yield analysis.

RESEARCH TREATMENTS:

BIOLOGICAL PRODUCT:

1. No biological
2. Envita SC foliar applied at V6 (Azotic NA)
3. UtrishaN foliar applied at V6 (Corteva Agrisciences)
4. Proven40OS seed treatment applied at planting (Pivot Bio)
5. Source foliar applied at V6 (Sound Ag)

NITROGEN FERTILIZER RATE:

1. 0 lbs N/ac
2. 60 lbs N/ac
3. 120 lbs N/ac
4. 180 lbs N/ac

TABLE 29. *Mean grain yield differences observed across biological products and nitrogen (N) fertilizer application rates in 2024. Wanatah, IN.*

NITROGEN RATE	BIOLOGICAL PRODUCT	GRAIN YIELD
-- lbs N/ac --		-- bu/ac --
0	None	168.6 h
	Envita SC	181.3 fgh
	Proven40 OS	179.3 gh
	Source	178.7 h
	Utrisha N	168.7 h
60	None	192.5 efg
	Envita SC	192.2 efg
	Proven40 OS	206.1 de
	Source	196.9 e
	Utrisha N	194.2 ef
120	None	218.1 bcd
	Envita SC	212.1 cd
	Proven40 OS	217.8 bcd
	Source	227.5 ab
	Utrisha N	221.3 abc
180	None	223.9 abc
	Envita SC	232.7 a
	Proven40 OS	226.6 ab
	Source	226.5 ab
	Utrisha N	229.2 ab

* Mean values that do not contain the same corresponding letter are determined statistically different ($P < 0.1$).

TABLE 30. *Corn grain yield (bu/ac) response to biological products. Data combined across all applied nitrogen fertilizer rates in the research trial. Wanatah, IN 2024.*

TREATMENT DESCRIPTION	GRAIN YIELD
	---- bu/ac ----
Nontreated	200.8 a
Source—Foliar Applied at V6	207.4 a
Envita SC—Foliar Applied at V6	204.5 a
Utrisha-N—Foliar Applied at V6	203.3 a
Proven 40 OS—Seed Trt	207.4 a

* Mean values which contain dissimilar letters and are in the same column are determined significantly different from each other ($P<0.1$).

SUMMARY (TAKE-HOME POINTS):

- Corn yield was significantly ($P < 0.1$) increased with increases in N fertilizer rates. However, preliminary results suggest examined biological products did not improve corn yield at any of the examined N rates (Table 29).
- Preliminary results suggest application of asymbiotic N-fixing bioinoculants do not improve corn yield or reduce synthetic N fertilizer requirements for corn at this location in 2024 (Tables 29 and 30).

POPCORN AND DENT CORN RESPONSE TO NITROGEN RATE (PPAC)

Narciso Zapata: Department of Agronomy, Purdue University
Daniel Quinn: Department of Agronomy, Purdue University
Stephen Boyer: Pinney Purdue Agricultural Center

Study Location: Pinney Purdue Agricultural Center, Wanatah, IN
Soil Type: Sebewa loam (0–2% slope)
Planting Date: May 23, 2024 | **Harvest Date:** October 22, 2024
Corn Hybrid: Weaver 2021 and Pioneer P1108Q | **Corn Seeding Rate:** 34,000 seeds/ac
Corn Nitrogen (N) Fertilizer Rate and Source: 0, 60, 120, 180, and 240 lbs N/ac. All nitrogen fertilizer rate applications were applied as 28% UAN coulter-injected as a sidedress application between the corn rows at growth stage V4. No starter fertilizer was used in this trial.
Previous Crop: Soybean | **Tillage:** Conventional
Study Replications: 5

RESEARCH TRIAL OVERVIEW:

A field research trial was established at the Pinney Purdue Agricultural Center (PPAC) in Porter County, IN. The research trial assessed corn yield differences when utilizing different N fertilizer application rates in popcorn and dent corn. The trial was designed as a randomized complete block design with 10 treatments and five replications. Plots measured 15 feet wide (six 30-inch corn rows) by 40 feet long and the center four rows were harvested with a small-plot combine and adjusted to 15.5% moisture for yield analysis.

RESEARCH TREATMENTS:

CORN TYPE:
1. Dent Corn
2. Popcorn

NITROGEN FERTILIZER RATE (COULTER-INJECTED BETWEEN THE CORN ROWS AT V5, NO STARTER APPLIED):
1. 0 lbs N/ac
2. 60 lbs N/ac
3. 120 lbs N/ac
4. 180 lbs N/ac
5. 240 lbs N/ac

RESULTS:

TABLE 31. *Quadratic plateau regression analysis parameters and calculated agronomic optimum nitrogen fertilizer rates (AONR) and yields obtained at the calculated AONR values for popcorn and dent corn. West Lafayette, IN 2024.*

| CORN TYPE | MODEL† | REGRESSION PARAMETERS | | | AONR | YAONR† |
		INTERCEPT	LINEAR COEFFICIENT	QUADRATIC COEFFICIENT		
					lbs N ac^{-1}	lbs ac^{-1}
Dent Corn	QP	154.3	0.61	-0.001	**225.8**	234.9
Popcorn	QP	4373.2	19.63	-0.054	**179.7**	6137.7

† QP, quadratic-plateau regression model; YAONR, popcorn grain yield at the agronomic optimum N rate (AONR).

SUMMARY (TAKE-HOME POINTS):

- Agronomic optimum N rate (AONR) for popcorn was 179.7 lbs N/ac to produce a grain yield of 6137 lbs/ac (Table 31). In comparison, the AONR for dent corn grown in the same environment, same field, and using the same management practices was 225.8 lbs N/ac to produce a maximum grain yield of 235 bu/ac.

- Results suggest at this location in 2024, popcorn required ~45 lbs N/ac less to produce maximum grain yield in comparison to dent corn.

DAVIS PURDUE AGRICULTURAL CENTER (DPAC)

CORN YIELD AND NITROGEN FERTILIZER RESPONSE TO ASYMBIOTIC N-FIXING BIOINOCULANT PRODUCTS (DPAC)

Daniel Quinn: Department of Agronomy, Purdue University
Jeff Boyer: Purdue Agricultural Center
Dwayne Boggs: Purdue Agricultural Center
Phil Tharp: Purdue Agricultural Center

Study Location: Davis Purdue Agricultural Center, Farmland, IN.
Soil Type: Blount Silt Loam (0–2% slope), Pewamo Silty Clay Loam (0–2% slope), Glynwood Clay Loam 6–12% slope), Glynwood Clay Loam (1–4% slope)
Planting Date: May 14, 2024 | **Harvest Date:** Oct. 30, 2024
Corn Hybrid: Pioneer 1108Q | **Corn Seeding Rate:** 32,000 seeds/ac
Corn Nitrogen (N) Fertilizer Rate and Source: 0, 60, 120, 180, and 240 lbs N/ac. All nitrogen fertilizer rate applications were applied as 28% UAN coulter-injected as a sidedress application between the corn rows at growth stage V4. No starter fertilizer was used in this trial.
Previous Crop: Soybeans | **Tillage:** Strip-till
Study Replications: 4

RESEARCH TRIAL OVERVIEW:

A field research trial was established at the Davis Purdue Agricultural Center (DPAC) in Randolph County, IN. The research trial examined corn yield response to Proven40OS applied as a seed treatment at planting and four different N fertilizer rates. The trial was designed as a randomized complete block design with four replications. Plot size measured 30 feet wide (12, 30-inch corn rows) x 600+ feet long. The center eight rows of each treatment were harvested using a commercial Case-IH combine and a calibrated AgLeader yield monitor. Grain yield values were adjusted to 15.5% moisture prior to statistical analysis.

RESEARCH TREATMENTS:

BIOLOGICAL PRODUCT:
1. No biological
2. Proven40OS seed treatment applied at planting (Pivot Bio)

NITROGEN FERTILIZER RATE (COULTER-INJECTED BETWEEN THE CORN ROWS AT V4, NO STARTER APPLIED):
1. 0 lbs N/ac
2. 60 lbs N/ac
3. 120 lbs N/ac
4. 180 lbs N/ac
5. 240 lbs N/ac

RESULTS:

TABLE 32. *Corn yield (bu/ac) interaction in response to seed-applied Proven40 OS and nitrogen fertilizer rate (lbs N/ac). Farmland, IN 2024.*

SEED-APPLIED PROVEN40	NITROGEN FERTILIZER RATE	GRAIN YIELD
	---- lbs N/ac ----	---- bu/ac ----
No	0	71.2 f*
Yes	0	77.7 f
No	60	129.8 e
Yes	60	131.9 e
No	120	201.5 d
Yes	120	207.3 cd
No	180	226.8 ab
Yes	180	213.2 bcd
No	240	234.4 a
Yes	240	226.8 ab

* Grain yield values adjusted to dry moisture of 15.5%. Mean yield values that contain a dissimilar letter are determined significantly different ($P < 0.1$).

TABLE 33. *Agronomic optimum nitrogen rate (AONR; nitrogen fertilizer rate required to maximize grain yield) differences observed between biological treatments. AONR values calculated by quadratic plateau regression analysis. Farmland, IN 2024.*

SEED-APPLIED PROVEN40	AGRONOMIC OPTIMUM NITROGEN RATE (AONR)
	----- lbs N/ac -----
Yes	206
No	239

SUMMARY (TAKE-HOME POINTS):

- The inclusion of Proven40OS reduced the agronomic optimum N rate at this location by ~33 lbs N/ac (Table 33). However, grain yield at the AONR and the higher N fertilizer rates applied was also reduced from inclusion of Proven40OS (Table 32).
- Overall, results suggest no additional benefit from the inclusion of Proven40OS at this location in 2024 despite the observed differences in N fertilizer rate response.

CORN YIELD RESPONSE TO AN INTEGRATED BIOSTIMULANT PROGRAM (DPAC)

Daniel Quinn: Department of Agronomy, Purdue University

Jeff Boyer: Purdue Agricultural Center

Dwayne Boggs: Purdue Agricultural Center

Phil Tharp: Purdue Agricultural Center

Study Location: Davis Purdue Agricultural Center, Farmland, IN

Soil Type: Blount Silt Loam (0–2% slope), Pewamo Silty Clay Loam (0–2% slope), Glynwood Clay Loam (1–4% slope)

Planting Date: May 14, 2024 | **Harvest Date:** Oct. 30, 2024

Corn Hybrid: Pioneer 1108Q | **Corn Seeding Rate:** 32,000 seeds/ac

Corn Nitrogen (N) Fertilizer Rate and Source: 240 lbs N/ac applied as 28% UAN coulter-injected as a sidedress application between the corn rows at growth stage V4.

Previous Crop: Soybeans | **Tillage:** Strip-till

Study Replications: 5

RESEARCH TRIAL OVERVIEW:

A field research trial was established at the Davis Purdue Agricultural Center (DPAC) in Randolph County, IN. The research trial examined corn yield response to a Pioneer Integrated Biostimulant Program. Fortified Stimulate Yield Enhancer Plus was applied both as an in-furrow application at planting and a V4 growth stage foliar spray. Sugar Mover Premier and X-Cyte were tank-mixed and applied as a foliar spray at the R1 growth stage. The trial was designed as a randomized complete block design with five replications. Plot size measured 30 feet wide (12, 30-inch corn rows) x 600+ feet long. The center eight rows of each treatment were harvested using a commercial Case-IH combine and a calibrated AgLeader yield monitor. Grain yield values are adjusted to 15.5% moisture.

RESEARCH TREATMENTS:

TABLE 34. *Description of research treatments applied. Farmland, IN 2024.*

TREATMENT*	TREATMENT DESCRIPTION
1	Untreated
2	Fortified Stimulate Yield Enhancer Plus (4 oz/ac) + Sugar Mover Premier (32 oz/ac) + X-Cyte (8 oz/ac)

*Field-scale research trial with treatments randomized and replicated 5 times. Plots measured 12 rows wide x length of the field. Nitrogen fertilizer was applied as 28% UAN at sidedress.

RESULTS:

TABLE 35. *Corn yield (bu/ac) response to biostimulant integrated program. Farmland, IN 2024.*

TRIAL TREATMENT	GRAIN YIELD
	---- bu/ac ----
1	240.6 a*
2	236.7 a

* Grain yield values adjusted to dry moisture of 15.5%. Mean yield values that contain a dissimilar letter are determined significantly different ($P < 0.1$).

SUMMARY (TAKE-HOME POINTS):

- No statistical yield increases were observed across treatments in this research trial in comparison to the nontreated control (Table 35).

NORTHEAST PURDUE AGRICULTURAL CENTER (NEPAC)

CORN YIELD RESPONSE TO ASYMBIOTIC N-FIXING BIOINOCULANT PRODUCTS (NEPAC)

Daniel Quinn: Department of Agronomy, Purdue University
Chris Lake: Northeast Purdue Agricultural Center
Carl Emley: Northeast Purdue Agricultural Center

Study Location: Northeast Purdue Agricultural Center, Columbia City, IN
Soil Type: Rawson sandy loam (2–6% slope), Glynwood loam (2–6% slope)
Planting Date: May 29, 2024 | **Harvest Date:** Oct. 30, 2024.
Corn Hybrid: Pioneer 1108Q | **Corn Seeding Rate:** 32,000 seeds/ac
Corn Nitrogen (N) Fertilizer Rate and Source: 200 lbs N/ac applied as 28% UAN coulter-injected as a sidedress application between the corn rows at growth stage V4. No starter fertilizer was applied in this research trial.
Previous Crop: Soybean | **Tillage:** No-till
Study Replications: 4

RESEARCH TRIAL OVERVIEW:

A field research trial was established at the Northeast Purdue Agricultural Center in Whitely County, IN. The research trial examined corn yield response to different biological products and N rates to analyze the impact of the interaction between both factors. Biological products Envita SC and Utrisha N were foliar applied on the same date at the V6 corn growth stage with a commercial sprayer with a carrier spray volume of 15 gallons per acre. Proven40OS was applied as a seed treatment at planting. The trial was designed as a randomized complete block design with four replications. Plot size measured 30 feet wide (12, 30-inch corn rows) x 400+ feet long. The center eight rows were harvested using a commercial Case-IH combine and a calibrated AgLeader yield monitor. Grain yield values are adjusted to 15.5% moisture.

RESEARCH TREATMENTS:

BIOLOGICAL PRODUCT:

1. No Biological
2. Envita SC foliar applied at V6 (Azotic NA)
3. Utrisha N foliar applied at V6 (Corteva Agrisciences)
4. Proven40OS seed treatment applied at planting (Pivot Bio)

RESULTS:

TABLE 36. *Corn grain yield (bu/ac) response to biological products. Columbia City, IN 2024.*

TREATMENT DESCRIPTION	GRAIN YIELD
	---- bu/ac ----
Nontreated Check	192.5 b
Envita SC—Foliar Applied at V6	200.7 a
Utrisha-N—Foliar Applied at V6	192.5 b
Proven40OS—Seed Trt	196.5 ab

* Mean values that contain dissimilar letters and are in the same column are determined significantly different from each other ($P < 0.1$).

SUMMARY (TAKE-HOME POINTS):

- The inclusion of Envita SC foliar applied at the V6 growth stage increased corn yield by 8 bu/ac in comparison to the nontreated control (Table 36).
- The research trial will be expanded and repeated in 2025 for further examination.

SOUTHEAST PURDUE AGRICULTURAL CENTER (SEPAC)

CORN RESPONSE TO IN-FURROW ENZUP K DS ACROSS DIFFERENT SOIL TEST K LEVELS (SEPAC)

Daniel Quinn: Department of Agronomy, Purdue University
Joel Wahlman: Southeast Purdue Agricultural Center (SEPAC)
Alex Helms: Southeast Purdue Agricultural Center (SEPAC)

Study Location: Southeast Purdue Agricultural Center, Butlerville, IN
Soil Type: Cobbsfork silt loam (0–1% slope)
Planting Date: May 24, 2024 | **Harvest Date:** Oct. 29, 2024
Corn Hybrid: Pioneer P1222AM | **Corn Seeding Rate:** 32,000 seeds/ac
Corn Nitrogen (N) Fertilizer Rate and Source: 40 lbs/ac applied as 28% UAN in a 2x2 starter at planting + 170 lbs N/ac applied as 28% UAN and coulter-injected as a sidedress application at the V3/V4 growth stage.
Previous Crop: Soybean | **Tillage:** No-till
Study Replications: 4

RESEARCH TRIAL OVERVIEW:

A field research trial was conducted at the Southeast Purdue Agricultural Center (SEPAC) in Jennings County, IN. The research trial examined corn yield response to in-furrow EnzUP K DS applied at a rate of 7 lbs/ac (49% soluble potash) dissolved in water across different soil test K levels and different spring broadcast-applied potash (0-0-61) rates. The trial was designed as a split-plot, randomized complete block design with 10 treatments and four replications. Plots measured 30 feet wide (12, 30-inch rows) by 150 feet long and the center four rows were harvested using a Wintersteiger Split Plot Combine equipped with a Harvest-Master weigh system and adjusted to 15.5% moisture.

RESEARCH TREATMENTS:

1. No potash (0-0-61) applied
2. No potash applied plus in-furrow EnzUp K DS at 7 lbs/ac (49% soluble potash)

3. 75 lbs/ac potash applied
4. 75 lbs/ac potash plus in-furrow EnzUpK DS at 7 lbs/ac
5. 150 lbs/ac potash applied
6. 150 lbs/ac potash plus in-furrow EnzUp K DS at 7 lbs/ac
7. 225 lbs/ac potash applied
8. 225 lbs/ac potash plus in-furrow EnzUp K DS at 7 lbs/ac
9. 300 lbs/ac potash applied
10. 300 lbs/ac potash plus in-furrow Enz Up K DS at 7 lbs/ac

RESULTS:

TABLE 37. *Corn grain yield response to in-furrow EnzUp K DS application at low spring soil test K values (STK) and with and without different spring broadcast rates of Potash (0-0-61). Butlerville, IN 2024.*

2024 POTASH RATE	2024 SPRING STK†	IN-FURROW ENZUP K APPLIED?‡	GRAIN YIELD
-- lbs/ac --	-- ppm --		-- bu/ac --
0	26.1	No	33.2 g*
		Yes	52.7 f
75	27.0	No	147.6 e
		Yes	175.6 d
150	26.6	No	221.6 c
		Yes	228.8 bc
225	31.1	No	237.5 ab
		Yes	243.5 a
300	29.4	No	232.3 bc
		Yes	246.3 a

† Spring 2024 soil samples (0–8") were sampled prior to spring broadcast potash application.
‡ EnzUp K was applied as an in-furrow application at planting and was mixed with water and applied at a rate equivalent to 7 lbs/ac (49% soluble potash).
* Mean grain yield values that do not contain the same corresponding letter are determined statistically different ($P < 0.1$).

SUMMARY (TAKE-HOME POINTS):

- Across all spring soil test K values and broadcast potash rates, the inclusion of EnzUp K DS as an in-furrow application at planting increased corn yield by 15 bu/ac (Table 37).
- The largest response to the in-furrow application of EnzUp K was observed at the lowest spring broadcast potash rates (0 and 75 lbs/ac) with yield increases of 19 and 28 bu/ac, respectively, compared to no in-furrow application (Table 37).
- Preliminary results suggest an in-furrow K application may be beneficial and improve yield beyond a spring potash application in low soil test K environments. However, this research trial will be repeated and expanded across more locations in 2025.

COMPARISON OF SHORT- AND TALL-STATURE CORN HYBRIDS TO NITROGEN FERTILIZER AND SEEDING RATES (SEPAC)

Erick Oliva: Department of Agronomy, Purdue University

Joel Wahlman: Southeast Purdue Agricultural Center (SEPAC)

Alex Helms: Southeast Purdue Agricultural Center (SEPAC)

Steven Ricker: Southeast Purdue Agricultural Center (SEPAC)

Daniel Quinn: Department of Agronomy, Purdue University

Study Location: Southeast Purdue Agricultural Center, Butlerville, IN.

Soil Type: Cobbsfork silt loam (0–1% slope)

Planting Date: May 21, 2024 | **Harvest Date:** October 28, 2024

Corn Hybrid: Preceon PR111-20SSC, Precion PR112-20SSC, and Dekalb DKC61-41RIB, Dekalb DKC62-70RIB

Corn Seeding Rate: 32,000, 38,000 and 44,000 seeds/ac

Corn Nitrogen (N) Fertilizer Rate and Source: 0, 90, 180 and 270 lbs N/ac as

28% UAN and applied as a coulter-injected sidedress application at the V5 growth stage. No starter fertilizer N was used in this trial.

Previous Crop: Soybean | **Tillage:** No-till

Study Replications: 4

RESEARCH TRIAL OVERVIEW:

A field research trial was established at the Southeast Purdue Agricultural Center (SEPAC) in Jennings County, IN. The trial examined corn yield response to different short and full-size corn hybrids, nitrogen, and seeding rates. The trial was designed as a split-plot, randomized complete block design with 48 treatments and four replications. Plots measured 15 feet wide (six 30-inch corn rows) by 100 feet long and the center four rows were harvested using a Wintersteiger plot combine equipped with a HarvestMaster weigh system and adjusted to 15.5% moisture for yield analysis.

RESULTS:

TABLE 38. *Short-stature (Preceon) and tall-stature (Dekalb) corn yield (bu/ac) and lodging rate (%) in response to hybrid type and seeding rate. Butlerville, IN 2024.*

TARGETED SEED RATE	HYBRID (TYPE)	GRAIN YIELD	LODGING RATE
		-- bu/ac --	--%--
32,000 seeds/ac	PR111-20SSC (Short)	194 b	0.0 a*
	PR116-20SSC (Short)	186 b	0.0 a
	DKC61-41RIB (Tall)	222 a	0.5 a
	DKC62-70RIB (Tall)	228 a	13.0 b
38,000 seeds/ac	PR111-20SSC (Short)	206 b	0.0 a
	PR116-20SSC (Short)	203 b	0.0 a
	DKC61-41RIB (Tall)	235 a	2.0 a
	DKC62-70RIB (Tall)	216 b	21.0 b
44,000 seeds/ac	PR111-20SSC (Short)	218 ab	0.0 a
	PR116-20SSC (Short)	205 b	0.0 a
	DKC61-41RIB (Tall)	234 a	3.4 a
	DKC62-70RIB (Tall)	208 b	41.0 b

* Mean values that do not contain the same corresponding letter are determined statistically different ($P < 0.1$). Columns with mean values that do not contain any letters are determined as no statistical differences between treatments.

TABLE 39. *Short-stature and tall-stature corn grain moisture and yield in response to hybrid and nitrogen rate. Butlerville, IN 2024.*

HYBRID	NITROGEN RATE	GRAIN MOISTURE	GRAIN YIELD
	--lbs N/ac--	-- % --	-- bu/ac --
PR111-20SSC	0	15.8 c	134 c*
	90	18.6 b	211 b
	180	20.4 a	242 a
	270	20.1 a	236 a
PR116-20SSC	0	19.6 c	154 c
	90	20.1 bc	192 b
	180	21.3 ab	226 a
	270	21.7 a	219 a
DKC61-41RIB	0	16.0 c	169 c
	90	17.2 bc	230 b
	180	17.6 ab	255 a
	270	18.8 a	268 a
DKC62-70RIB	0	16.3 b	165 b
	90	17.6 ab	234 a
	180	18.1 a	230 a
	270	18.4 a	239 a

* Mean values that do not contain the same corresponding letter are determined statistically different ($P < 0.1$). Columns with mean values that do not contain any letters are determined as no statistical differences between treatments.

SUMMARY (TAKE-HOME POINTS):

- A significant hybrid x seeding rate interaction was observed in this research trial (Table 38), which suggests that the different hybrids examined had different optimum seeding rates in this year and environment.
- Across all hybrids examined, a significant interaction between hybrid and N rate at SEPAC was observed in 2024, likely due to hybrid lodging differences (Table 39).
- Tall-stature hybrid DKC62-70RIB resulted in significant differences in lodging rate (Table 39) due to wind gusts of 36 mph and average wind events of 17 mph, remnants from Hurricane Helene.

CORN YIELD RESPONSE FOLIAR FUNGICIDE AND FOLIAR FERTILIZER APPLICATIONS AT THE R1 GROWTH STAGE (SEPAC)

Daniel Quinn: Department of Agronomy, Purdue University

Joel Wahlman: Southeast Purdue Agricultural Center (SEPAC)

Alex Helms: Southeast Purdue Agricultural Center (SEPAC)

Study Location: Southeast Purdue Agricultural Center, Butlerville, IN

Soil Type: Cobbsfork silt loam (0–1% slope)

Planting Date: May 24, 2024 | **Harvest Date:** Oct. 29, 2024

Corn Hybrid: Pioneer P1135AM | **Corn Seeding Rate:** 32,000 seeds/ac

Corn Nitrogen (N) Fertilizer Rate and Source: 40 lbs N/ac applied as 28% UAN in a 2x2 starter at planting + 170 lbs N/ac applied as 28% UAN and coulter-injected as a sidedress application at the V3/V4 growth stage.

Previous Crop: Soybean | **Tillage:** No-till

Study Replications: 5

RESEARCH TRIAL OVERVIEW:

A field research trial was conducted at the Southeast Purdue Agricultural Center (SEPAC) in Jennings County, IN. The research trial examined corn yield response to Foliar Smart KB and foliar fungicide applied at the R1 growth stage. The trial was designed as a randomized complete block design with four treatments and five replications. Plots measured 30 feet wide (12, 30-inch rows) by 700 feet long and the center six rows were harvested using a commercial John Deere combine with an AgLeader calibrated yield monitor and adjusted to 15.5% moisture prior to analysis.

RESEARCH TREATMENTS:

1. Nontreated Control
2. Delaro Complete (10 oz/ac at R1)
3. Delaro Complete (10 oz/ac) + Smart KB (2 qt/ac) at R1

RESULTS:

TABLE 40. *Corn grain moisture (%) and yield (bu/ac) in response to R1 foliar-applied fungicide and fertilizer products. Butlerville, IN 2024.*

FOLIAR TREATMENT	GRAIN MOISTURE	GRAIN YIELD
	---- % ----	---- bu/ac ----
Nontreated	23.5 a	249.2 a
Delaro Complete (10 oz/ac @ R1)	23.9 a	250.9 a
Delaro Complete + SmartKB	23.5 a	251.3 a

* Mean values that do not contain the same letter are determined statistically different from each other ($P < 0.1$).

SUMMARY (TAKE-HOME POINTS):

- No statistical yield increases were observed across treatments in this research trial in comparison to the nontreated control (Table 40). Research trial exhibited minimal foliar disease presence and/or nutrient deficiencies.

CORN RESPONSE TO FLUTRIAFOL FUNGICIDE APPLICATION TIMING AND PLACEMENT (SEPAC)

Daniel Quinn: Department of Agronomy, Purdue University

Joel Wahlman: Southeast Purdue Agricultural Center (SEPAC)

Alex Helms: Southeast Purdue Agricultural Center (SEPAC)

Study Location: Southeast Purdue Agricultural Center, Butlerville, IN

Soil Type: Cobbsfork silt loam (0-1% slope)

Planting Date: May 26, 2024 | **Harvest Date:** Oct. 29, 2024

Corn Hybrid: Pioneer 1136AM | **Corn Seeding Rate:** 32,000 seeds/ac

Corn Nitrogen (N) Fertilizer Rate and Source: 40 lbs N/ac as 28% UAN in a 2x2 starter + 170 lbs N/ac as 28% UAN either coulter-injected as a sidedress application between the corn rows or surface-banded next to the corn rows at growth stage V5. Sidedress N application method was dependent on required Xyway LFR application method.

Previous Crop: Soybean | **Tillage:** No-till

Study Replications: 5

RESEARCH TRIAL OVERVIEW:

A field research trial was established at the Southeast Purdue Agricultural Center (SEPAC) in Jennings County, IN. The research trial examined corn yield response to Xyway LFR fungicide applied with three different application methods at a rate of 15.2 oz/ac. These application methods included: (1) subsurface band (2x2) applied at planting, (2) coulter-injected band next to the corn row at the V6 growth stage (NR; next to row), and (3) coulter-injected band between the corn rows (BR; between rows) at the V6 growth stage compared to a nontreated check. In all application methods, Xyway LFR was tank-mixed and applied with 28% UAN fertilizer. The trial was designed as a randomized complete block design with four treatments and five replications. Plots measured 30 feet wide (12, 30-inch rows) by 1000 feet long. Grain yield was harvested from the center six rows using a commercial John Deere combine with a calibrated yield monitor (AgLeader) and adjusted to 15.5% moisture.

RESEARCH TREATMENTS:

1. Nontreated control (no fungicide applied)
2. Xyway LFR (15 oz/ac) applied as a 2x2 starter application at planting
3. Xyway LFR (15 oz/ac) surface-banded next to the row (NR; next to row) at the V6 growth stage
4. Xyway LFR (15 oz/ac) coulter-injected band between the corn rows (BR; between rows) at the V6 growth stage

RESULTS:

TABLE 41. *Corn grain moisture (%) and yield (bu/ac) in response to Xyway LFR placement. Butlerville, IN 2024.*

XYWAY LFR PLACEMENT	GRAIN MOISTURE	GRAIN YIELD
	---- % ----	---- bu/ac ----
Nontreated	22.1 a*	257.3 a
2x2 (At-planting)	22.1 a	259.9 a
NR (Next to row @ V6)	21.4 b	261.5 a
BR (Between rows @ V6)	21.8 a	257.1 a
Pr > F	0.0100	0.4678

* Mean values that do not contain the same letter are determined statistically different from each other ($P < 0.1$).

TABLE 42. *Corn ear leaf disease severity (R5 growth stage) in response to Xyway LFR placement. Butlerville, IN 2024.*

XYWAY LFR PLACEMENT	R5 EAR LF	R5 EAR LF
	--- % tar spot ---	--- % gray leaf spot ---
Nontreated	3.2 a*	3.8 a
2x2 (At-planting)	2.0 b	1.2 c
NR (Next to row @ V6)	1.4 b	2.2 bc
BR (Between rows @ V6)	3.2 a	2.8 ab
Pr > F	0.0119	0.0928

* Mean values that do not contain the same letter are determined statistically different from each other ($P < 0.1$).

SUMMARY (TAKE-HOME POINTS):

- Corn grain yield was not improved with the inclusion of Xyway LFR in comparison to the nontreated control and across various application methods (Table 41).
- Corn ear leaf disease severity at the R5 growth stage was reduced with the 2x2 and NR application methods in comparison to the nontreated control (Table 42). However, disease severity levels were too low to expect a corresponding yield response.

THROCKMORTON PURDUE AGRICULTURAL CENTER (TPAC)

CORN RESPONSE TO KORN-KALI AND KORN-KALI + B AT VARIOUS K RATES AND FERTILIZER COMBINATIONS (TPAC)

Daniel Quinn: Department of Agronomy, Purdue University
Jay Young: Throckmorton Purdue Agricultural Center (TPAC)
Pete Illingsworth: Throckmorton Purdue Agricultural Center (TPAC)

Study Location: Throckmorton Purdue Agricultural Center, Lafayette, IN
Soil Type: Toronto-Millbrook complex (0–2%), Octagon silt loam, eroded (2–6% slope)
Planting Date: May 21, 2024 | **Harvest Date:** Oct. 15, 2024.
Corn Hybrid: Pioneer P1108Q | **Corn Seeding Rate:** 30,000 seeds/ac
Corn Nitrogen (N) Fertilizer Rate and Source: 40 lbs N/ac as 28% UAN in a 2x2 starter at planting + 160 lbs N/ac as 28% UAN coulter-injected as a sidedress application between the corn rows at growth stage V5.
Previous Crop: Soybean | **Tillage:** Conventional Tillage
Study Replications: 5

RESEARCH TRIAL OVERVIEW:

A field research trial was established at the Throckmorton Purdue Agricultural Center (TPAC) in Lafayette, IN. The trial examined Korn KALI (0-0-40-6-13S) and Korn KALI +B (0-0-40-6-13S-0.25 B) with and without other soil K and S fertilizers. The dry fertilizer treatments were broadcast applied (1 day before planting) and incorporated. All combinations of K fertilizer products were applied to an equivalent rate of 100 lbs K/ac. The trial was designed as a randomized complete block design with seven treatments and five replications. Plot size measured 15 feet wide (six 30-inch corn rows) x 40 feet long and the center two rows were harvested using a Wintersteiger Plot Combine equipped with a HarvestMaster weigh system and adjusted to 15.5% moisture.

RESEARCH TREATMENTS:

1. Nontreated Control
2. Potash (0-0-60) applied at a rate equivalent to 100 lbs K/ac
3. Korn KALI (0-0-40-6-13S) at 50 lbs K/ac + Potash (0-0-60) at 50 lbs K/ac
4. Korn KALI +B (0-0-40-6-13S-0.25B) at 50 lbs K/ac + Potash (0-0-60) at 50 lbs K/ac
5. Korn KALI (0-0-40-6-13S) at 100 lbs K/ac
6. Korn KALI +B (0-0-40-6-13S-0.25B) at 100 lbs K/ac
7. K-Mag Standard (0-0-22-11-22S) at 59 lbs K/ac + Potash (0-0-60) at 41 lbs K/ac

RESULTS:

TABLE 43. *Corn plant stand (plants/ac), grain moisture (%), and yield (bu/ac) in response to fertilizer treatment. Lafayette, IN 2024.*

FERTILIZER TREATMENT	PLANT STAND	GRAIN MOISTURE	GRAIN YIELD
	-- plants/ac --	---- % ----	---- bu/ac ----
1. Nontreated (0 lbs K/ac)	29229 a*	20.1 a	242.8 a
2. Potash (0-0-60, 100 lbs K/ac)	29229 a	20.4 a	245.4 a
3. Korn-KALI (0-0-40-13S, 50 lbs K/ac) + Potash (50 lbs K/ac)	28924 a	20.4 a	244.2 a
4. Korn-KALI+B (0-0-40-13S, 50 lbs K/ac) + Potash (50 lbs K/ac)	28619 a	19.7 a	245.1 a
5. Korn-KALI (100 lbs K/ac)	28801 a	20.1 a	240.6 a
6. Korn-KALI + B (100 lbs K/ac)	28924 a	19.8 a	242.4 a
7. KMag (0-0-22-22S, 59 lbs K/ac) + Potash (41 lbs K/ac)	29098 a	20.2 a	246.8 a
Pr > F	0.7842	0.9529	0.9271

* Mean values that do not contain the same letter are determined statistically different from each other ($P < 0.1$).

SUMMARY (TAKE-HOME POINTS):

- No statistical plant stand, moisture, or yield differences were observed across treatments in this research trial (Table 43).
- Soil test potassium levels were all above critical at this research trial location.

CORN RESPONSE TO FLUTRIAFOL FUNGICIDE APPLICATION TIMING AND PLACEMENT (TPAC)

Daniel Quinn: Department of Agronomy, Purdue University

Jay Young: Throckmorton Purdue Agricultural Center (TPAC)

Pete Illingsworth: Throckmorton Purdue Agricultural Center (TPAC)

Study Location: Throckmorton Purdue Agricultural Center, Lafayette, IN

Soil Type: Drummer silty clay loam (0–2% slope)

Planting Date: May 21, 2024 | **Harvest Date:** Oct. 15, 2024

Corn Hybrid: Pioneer1108Q | **Corn Seeding Rate:** 32,000 seeds/ac

Corn Nitrogen (N) Fertilizer Rate and Source: 40 lbs N/ac as 28% UAN in a 2x2 starter + 160 lbs N/ac as 28% UAN either coulter-injected as a sidedress application between the corn rows or surface-banded next to the corn rows at growth stage V5. Sidedress N application method was dependent on required Xyway LFR application method.

Previous Crop: Soybean | **Tillage:** Spring Vertical Tillage

Study Replications: 5

RESEARCH TRIAL OVERVIEW:

A field research trial was established at the Throckmorton Purdue Agricultural Center (TPAC) in Lafayette, IN. The research trial examined corn yield response to Xyway LFR fungicide applied with three different application methods at a rate of 15.2 oz/ac. These application methods included: (1) subsurface band (2x2) applied at planting, (2) coulter-injected band next to the corn row at the V6 growth stage (NR; next to row), and (3) coulter-injected band in between the corn rows (BR; between rows) at the V6 growth stage compared to a nontreated check. In all application methods, Xyway LFR was tank-mixed and applied with 28% UAN fertilizer. The trial was designed as a randomized complete block design with four treatments and five replications. Plots measured 15 feet wide (six 30-inch rows) by 40 feet long. Grain yield was harvested from the center two rows using a Wintersteiger Plot Combine and adjusted to 15.5% moisture.

RESEARCH TREATMENTS:

1. Nontreated control (no fungicide applied)
2. Xyway LFR (15 oz/ac) applied as a 2x2 starter application at planting
3. Xyway LFR (15 oz/ac) surface-banded next to the row (NR; next to row) at the V6 growth stage
4. Xyway LFR (15 oz/ac) coulter-injected band between the corn rows (BR; between rows) at the V6 growth stage

RESULTS:

TABLE 44. *Corn grain moisture (%) and yield (bu/ac) in response to Xyway LFR placement. Lafayette, IN 2024.*

XYWAY LFR PLACEMENT	GRAIN MOISTURE	GRAIN YIELD
	---- % ----	---- bu/ac ----
Nontreated	20.1 a*	233.6 b
2x2 (At-planting)	20.3 a	251.1 a
NR (Next to row @ V6)	20.5 a	247.1 a
BR (Between rows @ V6)	19.9 a	249.8 a
Pr > F	0.7237	*0.0618*

* Mean values that do not contain the same letter are determined statistically different from each other (*P* < 0.1).

TABLE 45. *Corn ear leaf disease severity (R4 and R5 growth stage) in response to Xyway LFR placement. Lafayette, IN 2024.*

XYWAY LFR PLACEMENT	R4 EAR LF	R5 EAR LF
	- % tar spot -	- % tar spot -
Nontreated	36.6 a*	49.5 a
2x2 (At-planting)	20.1 b	27.8 b
NR (Next to row @ V6)	24.0 b	31.7 b
BR (Between rows @ V6)	22.5 b	30.9 b
Pr > F	*0.0676*	*0.0692*

* Mean values that do not contain the same letter are determined statistically different from each other (*P* < 0.1).

SUMMARY (TAKE-HOME POINTS):

- All flutriafol treatments examined reduced growth stage R4 and R5 ear leaf tar spot disease severity (Table 45).
- The inclusion of Xyway LFR, regardless of application method, improved corn grain yield in comparison to the nontreated control by an average of 16 bu/ac (Table 44). This result was likely due to the high disease pressure levels observed at this location in 2024.
- Preliminary results suggest that Xyway LFR applied as a sidedress application (NR or BR) was just as effective as the 2x2 application at this location in 2024.

CORN YIELD AND NITROGEN FERTILIZER RESPONSE TO ASYMBIOTIC N-FIXING BIOINOCULANT PRODUCTS (TPAC)

Daniel Quinn: Department of Agronomy, Purdue University

Jay Young: Throckmorton Purdue Agricultural Center (TPAC)

Pete Illingsworth: Throckmorton Purdue Agricultural Center (TPAC)

Study Location: Throckmorton Purdue Agricultural Center, Lafayette, IN

Soil Type: Starks-Fincastle complex (0–2% slope), Drummer (0–2% slope), Richlandville Silt Loam (0–2% slope), Toronto Millbrook complex (0–2% slope), Lauramie silt loam (0–2% slope)

Planting Date: May 26, 2024 | **Harvest Date:** Oct 21, 2024

Corn Hybrid: Pioneer 1108Q | **Corn Seeding Rate:** 32,000 seeds/ac

Corn Nitrogen (N) Fertilizer Rate and Source: 0, 100 and 200 lbs N/ac as 28% UAN coulter-injected as a sidedress application between the corn rows at growth stage V5. No starter fertilizer was included in this trial.

Previous Crop: Soybeans | **Tillage:** Conventional

Study Replications: 5

RESEARCH TRIAL OVERVIEW:

A field research trial was established at the Throckmorton Purdue Agricultural Center (TPAC) in Lafayette, IN. The research trial examined corn yield response to different biological products and N fertilizer rates to analyze the impact of the interaction between both factors. The trial was designed as a randomized complete block design with six treatments and five replications. Plot size measured 60 feet wide (24, 30-inch corn rows) x 1000+ feet long. The center 12 rows of each plot were harvested using a commercial Case-IH combine and a calibrated AgLeader yield monitor. Grain yield values are adjusted to 15.5% moisture prior to analysis.

RESEARCH TREATMENTS:

BIOLOGICAL PRODUCT:
1. No biological
2. Envita SC foliar applied at V6 (Azotic NA)
3. Source foliar applied at V6 (Sound Ag)

NITROGEN FERTILIZER RATE (COULTER-INJECTED BETWEEN THE CORN ROWS AT V5, NO STARTER APPLIED):
1. 0 lbs N/ac
2. 150 lbs N/ac
3. 200 lbs N/ac

RESULTS:

TABLE 46. *Mean grain yield (bu/ac) differences observed across biological products and nitrogen (N) fertilizer application rate in 2024. Lafayette, IN.*

NITROGEN RATE	BIOLOGICAL PRODUCT	GRAIN YIELD
-- lbs N/ac --		-- bu/ac --
150	None	234.6 a*
	EnvitaSC	224.3 b
	Source	230.9 ab
200	None	239.9 a
	EnvitaSC	231.6 ab
	Source	235.3 a

* Mean values that do not contain the same corresponding letter are determined statistically different ($P < 0.1$).

SUMMARY (TAKE-HOME POINTS):

- At the N fertilizer rate of 150 lbs N/ac, the inclusion of Envita SC statistically reduced corn grain yield by 10 bu/ac (Table 46).
- At the N fertilizer rate of 200 lbs N/ac, the inclusion of Envita SC and Source did not improve corn grain yield beyond the nontreated control.
- Preliminary results suggest no yield or N fertilizer response benefit from the inclusion of Envita SC or Source at this location in 2024.

WEATHER DATA

TABLE 47. *Growing season monthly total precipitation, average temperature, and their respective 30-yr (1991–2021) averages.* **West Lafayette, IN 2024 (ACRE).**

MONTH	TOTAL PRECIP	30-YR AVE PRECIP	AVERAGE TEMP	30-YR AVE TEMP
	--- in ---	--- in ---	--- °F ---	--- °F ---
April	5.85	3.84	54.1	51.1
May	2.14	4.04	67.3	61.6
June	2.80	4.56	73.7	70.7
July	5.01	4.08	72.7	73.6
August	4.15	3.12	71.8	72.2
September	1.42	2.59	67.0	65.5
October	0.19	2.91	57.3	53.7

TABLE 48. *Growing season monthly total precipitation, average temperature, and their respective 30-yr (1991–2021) averages.* **Lafayette, IN 2024 (TPAC).**

MONTH	TOTAL PRECIP	30-YR AVE PRECIP	AVERAGE TEMP	30-YR AVE TEMP
	--- in ---	--- in ---	--- °F ---	--- °F ---
April	5.63	3.84	54.7	51.1
May	2.90	4.04	68.1	61.6
June	9.68	4.56	74.4	70.7
July	5.28	4.08	73.5	73.6
August	2.59	3.12	73.1	72.2
September	1.14	2.59	68.8	65.5
October	0.35	2.91	59.1	53.7

TABLE 49. *Growing season monthly total precipitation, average temperature, and their respective 30-yr (1991–2021) averages.* **Wanatah, IN 2024 (PPAC).**

MONTH	TOTAL PRECIP	30-YR AVE PRECIP	AVERAGE TEMP	30-YR AVE TEMP
	--- in ---	--- in ---	--- °F ---	--- °F ---
April	4.65	3.78	50.7	48.1
May	2.86	4.00	63.0	59.9
June	2.34	4.71	71.3	69.7
July	6.16	4.18	70.2	72.4
August	2.95	4.38	69.7	70.3
September	2.39	3.28	64.9	63.7
October	1.86	3.78	55.0	52.0

TABLE 50. *Growing season monthly total precipitation, average temperature, and their respective 30-yr (1991–2021) averages.* **Columbia City, IN 2024 (NEPAC).**

MONTH	TOTAL PRECIP	30-YR AVE PRECIP	AVERAGE TEMP	30-YR AVE TEMP
	--- in ---	--- in ---	--- °F ---	--- °F ---
April	7.96	3.84	52.3	51.1
May	3.61	4.04	65.3	61.6
June	3.06	4.56	72.8	70.7
July	4.10	4.08	72.2	73.6
August	1.67	3.12	71.3	72.2
September	1.43	2.59	67.0	65.5
October	0.42	2.91	56.4	53.7

TABLE 51. *Growing season monthly total precipitation, average temperature, and their respective 30-yr (1991–2021) averages.* **Farmland, IN 2024 (DPAC).**

MONTH	TOTAL PRECIP	30-YR AVE PRECIP	AVERAGE TEMP	30-YR AVE TEMP
	--- in ---	--- in ---	--- °F ---	--- °F ---
April	6.29	3.84	54	51.1
May	1.78	4.04	66.5	61.6
June	2.43	4.56	72.7	70.7
July	2.22	4.08	71.9	73.6
August	3.83	3.12	71.1	72.2
September	2.20	2.59	66.3	65.5
October	0.22	2.91	56.0	53.7

TABLE 52. *Growing season monthly total precipitation, average temperature, and their respective 30-yr (1991–2021) averages.* **Butlerville, IN 2024 (SEPAC).**

MONTH	TOTAL PRECIP	30-YR AVE PRECIP	AVERAGE TEMP	30-YR AVE TEMP
	--- in ---	--- in ---	--- °F ---	--- °F ---
April	4.95	3.84	57.5	51.1
May	4.42	4.04	67.6	61.6
June	1.23	4.56	73.4	70.7
July	5.57	4.08	74.5	73.6
August	4.82	3.12	73.1	72.2
September	4.71	2.59	68.4	65.5
October	0.66	2.91	58.0	53.7

INTERESTED IN PARTICIPATING IN ON-FARM RESEARCH?

Interested in working with Purdue University to address management questions and improve your operation through on-farm research? Both the Purdue Corn Agronomy Team and the Purdue On The Farm Program continue to look for on-farm cooperators for participation and assistance with on-farm research trials. In addition, we will work closely with you to answer specific questions both we and you may have specific to your own operation. Information and data collected are shared directly to each cooperator every step of the way. For more information, please reach out directly below:

Dan Quinn, PhD
Extension Corn Specialist
Purdue University
Email: djquinn@purdue.edu

Betsy Bower
Senior Research Associate
Purdue University
Email: eabower@purdue.edu

Scott Gabbard
Purdue On The Farm Coordinator
Purdue University
Email: gabbardd@purdue.edu

ABOUT THE AUTHOR

DAN QUINN is an assistant professor of agronomy and an extension corn specialist at Purdue University. His research and extension program focuses on improving corn production systems in the Midwest through large-scale and small-plot field trials. Quinn's key areas of study include yield physiology, agronomic intensification, precision technologies, nutrient management, and cover crops, with an emphasis on enhancing profitability, productivity, and sustainability in corn-based agriculture.

www.ingramcontent.com/pod-product-compliance
Lightning Source LLC
Chambersburg PA
CBHW041449210326
41599CB00004B/188